Technology in Counselling

Also by Stephen Goss:

Evidence-based Counselling and the Psychological Therapies
(*with Nancy Rowland*)

Technology in Counselling and Psychotherapy

A Practitioner's Guide

edited by
STEPHEN GOSS
and
KATE ANTHONY

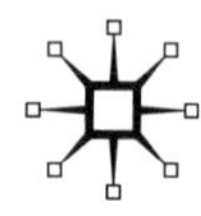

First published 2003 by
PALGRAVE MACMILLAN
Houndmills, Basingstoke, Hampshire RG21 6XS and
175 Fifth Avenue, New York, N.Y. 10010
Companies and representatives throughout the world

PALGRAVE MACMILLAN is the global academic imprint of the Palgrave Macmillan division of St. Martin's Press, LLC and of Palgrave Macmillan Ltd. Macmillan® is a registered trademark in the United States, United Kingdom and other countries. Palgrave is a registered trademark in the European Union and other countries.

ISBN 1–4039–0060–4 paperback

This book is printed on paper suitable for recycling and made from fully managed and sustained forest sources.

A catalogue record for this book is available from the British Library.

A catalog record for this book is available from the Library of Congress.

Editing and origination by Aardvark Editorial, Mendham, Suffolk

10 9 8 7 6 5 4 3 2 1
12 11 10 09 08 07 06 05 04 03

Printed and bound in Great Britain by
Creative Print & Design (Wales), Ebbw Vale

To Andrew, Catriona and Lynn and to Nick and Sidonie

Contents

List of screenshots

Notes on the contributors

Kate Anthony MSc runs www.OnlineCounsellors.co.uk which provides online and offline consultancy services and training for practitioners using the Internet. She conducts ongoing research programmes about online therapy and supervision, and is the author of some of the very few published empirical studies to be conducted into the use of email and Internet Relay Chat in therapy. She was a co-author of the BACP *Guidelines for Online Counselling and Psychotherapy* (BACP, 2001) and a regular contributor to journals, international conferences, media programmes and books. She is also a psychotherapist with Oxleas Health Trust in southeast London.

Phillip Armstrong B.Couns, Dip. Psych is the National Manager of the Australian Counselling Association and editor of the professional journal *Counselling Australia*. He is also Director of his own consultancy company which supplies counselling services and also runs workshops throughout Australia. He is a qualified supervisor and currently supervises 12 counsellors who are all in private practice. He has pioneered phone supervision in Australia over the last few years and has had several articles on professional supervision published. He is currently involved in redesigning and adding a completely new dynamic to employment assistant programmes in Australia.

Dr Kate Cavanagh is a Senior Psychologist at Ultrasis plc, an interactive healthcare company, and Hon. Researcher at the Institute of Psychiatry, Kings College, London. Her interests include the development and delivery of innovative health technologies including computerized psychotherapies. She is an Associate Editor of *The Psychologist* and has published peer-reviewed articles and book chapters as well as presenting her research around the world.

Peter J. Chechele is a licensed marriage and family therapist in the San Francisco Bay area. He began offering online therapy as an adjunct to his private practice in November of 1997 at www.cybertherapy.com. Over the past five years he has worked with several hundred clients online, continually developing his skills for working with clients through email, chat and telephone. Peter is a member of the board of directors for the International Society of Mental Health Online (www.ismho.org) and is one of the founding members of the millennium online clinical case study group.

Yvette Colón, MSW, is Director of Education and Internet services at the American Pain Foundation in Baltimore, MD, where she is responsible for educational programme development, including the implementation of online support groups and technology-based services as well as consultation/technical assistance to professionals and organizations nationally and internationally. She was recently awarded an Open Society Institute Project on Death in America Social Work Leadership Award to create a new interactive teaching model that utilizes the Internet to provide focused, intensive training on end-of-life issues to master's and post-master's level social work professionals. She has facilitated many online psychotherapy and cancer support groups since 1993. She teaches end-of-life social work practice at the Smith College School for Social Work in Northampton, MA, and is currently a PhD candidate at the New York University Ehrenkranz School of Social Work.

Dr Michael Fenichel is a Clinical Psychologist who practises in New York, provides supervision to practitioners and doctoral students and consults with media and business. His early research was in the area of 'person-therapy fit' and in recent years this interest has extended to a fascination with the potential and challenges of 'Internet-facilitated' relationships. He has written about this extensively, online and off, most recently in 'Myths and realities of online clinical work' (*CyberPsychology and Behavior* 5, 2002). He developed an award-winning psychology website, was a co-founder and past president of the International Society for Mental Health Online (ISMHO) and co-leader of ISMHO's clinical case study group since its inception.

Beth Friedman CSW, MPH is an oncology social worker and Programme Coordinator of online services at Cancer Care's New York office. She holds MAs in social work and public health from Columbia University. At Cancer Care, she provides psychotherapy to cancer patients and their families, individual and group psychotherapy with adults, bereavement therapy with family members, and conducts online support groups for patients, family members and the bereaved. She is responsible for the online services program development, including staff training, within Cancer Care, community education for patients, and consultation/technical assistance to professionals and organizations throughout the United States and worldwide. She has co-developed Cancer Care's online program and facilitated many online groups. She has written and presented extensively on developing and providing online services.

Dr Stephen Goss is Research Development Manager with the British Association for Counselling and Psychotherapy (BACP) and Hon. Research Fellow with the University of Strathclyde and a qualified counsellor and supervisor. His interests include innovative methods of service delivery in therapy as well as in research and evaluation methodologies, pluralist philosophies of science and maintenance of high practice standards. He was the lead author of the recent BACP *Guidelines for Online Counselling and Psychotherapy* (BACP, 2001). He has produced numerous research reports, journal articles and other works including the edited text *Evidence-based Counselling and Psychotherapy* with Nancy Rowland (Routledge, 2000).

Dr H. Lori Schnieders is Assistant Professor of Practice at Vanderbilt University in the Department of Human and Organizational Development, Division of Human Development Counseling. She is a qualified counsellor, supervisor and counsellor educator, specializing in working with children and adolescents in both a school and agency setting. Her research focus is on the empathic connection between children and parents, Internet use for cyber-supervision and multicultural counselling with children. During the 2001–02 academic year she worked with research partners in Scotland and Wales to connect cohorts of children via Internet-2. In addition to her research and writing, she has presented numerous workshops around the United States for teachers and counsellors.

Maxine Rosenfield is a counsellor, consultant and trainer with particular interest in working by telephone. She has written numerous articles and fact sheets about telephone counselling and telephone support work, including the BACP information sheet *Counselling by Telephone*, as well as writing the book *Counselling by Telephone* (Sage, 1997). Maxine was the first chair of the UK Telephone Helplines Association, having been involved in its development and the production of its publications such as good practice guidelines for helplines since 1991. In 1999 Maxine emigrated to Sydney, Australia and continues her work with helplines and call centres and writing articles and book chapters about telephone work. She is an assessor for the Quality Improvement Council programme of review and accreditation for community and health services and has a particular interest in the development, implementation and evaluation of good practice standards, benchmarks and performance indicators for helplines and call centres.

Dr David A. Shapiro is Hon. Professor at the University of Leeds and the University of Sheffield. A Clinical Psychologist, for many years he has researched comparative processes and outcomes of psychological treatments of depression. Other work has focused on occupational mental health. He has some 120 peer-reviewed publications. He led a research team at the former MRC/ESRC Social and Applied Psychology Unit from 1977 to 1994, before establishing the Psychological Therapies Research Centre at the University of Leeds in 1995. A former editor of the *British Journal of Clinical Psychology* and *Psychotherapy Research*, he served as President of the Society for Psychotherapy Research in 1993–4. He now works as an independent consultant offering research, evaluation and service development in clinical psychology, psychological therapies and occupational mental health. This includes working with Ultrasis plc on the effectiveness of its computerized, cognitive-behaviour therapy program, *Beating the Blues.*

Susan Simpson is a Clinical Psychologist working both in the NHS and the independent sector. She graduated in psychology from Adelaide University, and completed her clinical training at Flinders University of South Australia in 1994. She has worked in the UK for the past eight years, initially in Chichester and currently in Aberdeen. Her clinical interests include the treatment of people with personality disorders and chronic difficulties, and she has particular expertise in the treatment of eating disorders. She established a videoconferencing psychology service to the Shetland Isles, and extended the Grampian Eating Disorders Service to include remote and rural communities in the northeast of Scotland. She has published and lectured on the use of videoconferencing in therapy both in the UK and Norway and is currently conducting her doctoral dissertation research on the treatment of bulimic disorders via videoconferencing.

Gary Stofle LISW, CCDCIIIE is currently Team Leader for a dual diagnosis community treatment team in Columbus, Ohio. He has an ongoing interest in therapy conducted online and has written a book entitled *Choosing an Online Therapist: A Step by Step Guide to Finding Professional Help on the Web* (White Hat Communications, 2001) as well as several articles dealing with the provision of therapy services in a chat room. He has served as the Secretary/Treasurer of the International Society of Mental Health Online (ISMHO). He teaches and provides continuing education workshops online.

Dr Jesse H. Wright MD, PhD is a Professor in the Department of Psychiatry and Behavioral Sciences at the University of Louisville where he also serves as Associate Chairman and Chief of Adult Psychiatry. He is the principal author of the first multimedia computer program for cognitive therapy, *Good Days Ahead* (Wright et al., 2002), and the self-help book, *Getting Your Life Back: The Complete Guide to Recovery from Depression* (Free Press, 2001). A DVD-ROM version of his empirically tested software for computer-assisted cognitive therapy has been released recently. Dr Wright was the founding President of the Academy of Cognitive Therapy. He lectures widely on cognitive therapy, psychopharmacology and computer-aided psychotherapy.

Dr Jason S. Zack is President of JSZ Behavioral Science, Inc. and part-time faculty member at the University of Miami (Coral Gables, Florida, USA). He serves on the board of directors for the International Society for Mental Health Online and was formerly Director of research and development for eTherapy.com. An expert in psychotherapy research and psychotechnology, Dr Zack currently consults with a variety of organizations (primarily in the US and Japan) to develop new, technology-driven modes of mental health service delivery. He has published numerous articles in peer-review journals and has been invited to present his research around the world.

Preface

Technology in Counselling and Psychotherapy: A Practitioner's Guide is a very welcome addition to the practitioner's library. It comprehensively covers the whole range of technology that is available to practitioners: email, Internet Relay Chat, videoconferencing, telephone and stand-alone software. It explores not only the advantages and disadvantages of various applications but also clinical and ethical issues that arise from their use. As such, it is unique and ground-breaking, and fills a void that has existed for several years. There has been a growing demand from practitioners for an informative and substantive source of knowledge in this new specialism in our psychotherapeutic field.

This book addresses and fits well with a key objective of the British Association for Counselling and Psychotherapy (BACP): to encourage high standards in the practice and conduct of counsellors and psychotherapists. The two major methods by which the BACP reaches its objective are by supporting ongoing development of theory and practice and, in turn, supporting practitioners to keep abreast of new developments as part of their continuing professional development. I have no doubt that *Technology in Counselling and Psychotherapy* will help the BACP in the achievement of its objectives by creating solid ground not only for those practitioners already involved in using technology in their practice but also for those considering the possibilities this technology opens up. This book represents a quantum leap from where we were only five years ago. I am proud to see that BACP has played no small part in preparing the path for this book.

For a number of years, the BACP has been keeping a close eye on the development of online counselling and psychotherapy, both nationally and internationally. For me, this began around 1997 when, at the BACP annual conference in Edinburgh, I was asked if I would chair an impromptu discussion group on the viability of using technology as a means of delivering counselling. I was particularly keen to take part in this discussion because of a growing concern in my own counselling service at Edinburgh University. We were worried about an increasing number of students who were contacting their counsellors via email between sessions and during interruptions to sessions. We were finding that these students tended to make more and deeper disclosures about themselves in these emails than they seemed able or willing to do in actual counselling sessions. We were worried that email, with its absence of face-to-face encounter in real time, encouraged a dangerous degree of disinhibition. My team argued passionately about how one should respond. Were we encouraging something unethical? Should we discourage it?

In the discussion group in Edinburgh, the general drift of the discussion was that to attempt to offer therapy on the Internet was impractical, impossible and unethical. There was a rather complacent mood of 'it will never catch on' (I believe that the same was said of television back in the 1930s). My own view was coloured by a definite belief in humankind's infinite capacity to seize on something novel and in no time develop it into yet another of life's essentials. Amongst those in the discussion group was Stephen Goss, one of the editors of this book. I was pleased when he offered to keep a dialogue going with interested members of the group through email correspondence. Additionally, being a member of the BACP research committee, he brought the issue of online therapy onto the committee's agenda. Later, when BACP expanded its research department, he became our research and development manager. His appointment ensured online therapy would remain an important focus of the department's work and led to the BACP discussion document *Counselling Online ... Opportunities and Risks in Counselling Clients over the Internet* (BACP, 1999). At the same time, Kate Anthony, the co-editor of this book, was conducting one of the few empirical studies into the possibility of a therapeutic relationship existing over the Internet for her MSc thesis. The results of this study, and her fascination with how humans adapt to using technology to conduct a relationship, led her to a career in research, training and consultancy on the subject. Both editors' work ensured that online therapy was a recurrent feature of the BACP journal and, with co-authors Alan Jamieson and Professor Stephen Palmer, resulted in the development of perhaps the most definitive guidelines yet available, *Guidelines for Online Counselling and Psychotherapy* (BACP, 2001) which included the collaboration of an international network of practitioners involved in this specialist area. Several of those practitioners are contributors to *Technology in Counselling and Psychotherapy*.

This has all been a lot of work in a brief period of time, which seems par for the course in the rapid development of technology. The technology available to us today still has deficiencies in the interface and this book does not flinch or gloss over those difficulties. Despite these deficiencies, clients and therapists manage to find their way through this, work well and achieve good outcomes. It is only a matter of time until better technology will be developed and through economies of scale will become inexpensive and widely used. Somewhere down the road will be holographic projection so that client and therapist will be able to observe each other within their own three-dimensional physical space in real time. Exciting possibilities lie ahead.

Finally, I anticipate a side benefit through the publication of *Technology in Counselling and Psychotherapy*. The BACP is currently involved with other national associations in our professional field in exploring the possibility of the international adoption of protocols for the use and regulation of online therapy. The BACP and its fellow national and international associations have a responsibility to advocate regulation because usage of this technology is global and internationally recognized protocols are needed for the protection of the public. It is important to keep in mind that online therapy takes place across frontiers, time zones and beyond the jurisdiction of any particular professional, regulatory body. It is now easy for the depressed Los Angeles night owl to have online access to a bright-eyed, bushy-tailed, morning-type therapist in London or remote Stromness. Because *Technology in Counselling and Psychotherapy* is at the cutting edge and the result of international collaboration, it will inform and add weight to international discussions on the subject of regulation and the development of regulatory protocols.

CRAIG MCDEVITT
Chair of the British Association for Counselling and Psychotherapy

Acknowledgements

The editors would like to acknowledge the help and support during the development of this book of Alan Jamieson, Nick Bartlett, Bob Rich, Julie Monro, Geraldine Wilkes, Peter Smith, friends, family and the many colleagues too numerous to list.

Introduction

STEPHEN GOSS AND KATE ANTHONY

The idea of technology intervening in some way in the counselling or psychotherapeutic process seems to provoke strong reactions. In preparing this book, we have been aware of the absence of consensus within the profession on the technological applications that our contributors discuss. Some have welcomed unreservedly the idea of providing counselling and psychotherapy through technological media of one kind or another. For them, the idea seems like a natural means by which a client might contact a practitioner. They see few problems beyond translating their previous reliance on primarily verbal communication skills into the new context and, tentatively, explore what possibilities there may be. Some of our colleagues even express surprise and exasperation that it is not generally accepted already. Others, however, take the opposite position; without the full range of inflection and the subtle nuances in verbal and non-verbal communication it must surely be impossible or even dangerous, they suggest, to provide therapy when client and practitioner are many miles apart and perhaps never meet face to face at all. Some will dismiss the idea out of hand. Some are merely sceptical and wait to be convinced.

That views should be polarized in this way suggests that this is still a 'young' science. Taking the profession as a whole, we are not yet ready for that revolution in thought that renders a 'new' idea simply an accepted part of our world; for many, Douglas Adams' (1999) quote resonates:

> Another problem with the net is that it's still 'technology', and 'technology', as the computer scientist Bran Ferren memorably defined it, is 'stuff that doesn't work yet.' We no longer think of chairs as technology, we just think of them as chairs.

Currently, however, we must still consider the means of providing therapy presented in the following chapters to be at that stage that comes before an established paradigmatic view can be said to exist.

This is despite the relatively long history of the suggestion that technology might have much to offer the world of psychology and the psychological therapies. As early as the 1940s, the advantages of electronic recordings for analysis were noted by Carl Rogers (1942) and the 1960s saw several publications dealing with the possibilities of the then still emerging computer industry (for example Cogswell and Estavan, 1965; Weizenbaum, 1966; Cogswell et al., 1967). Perhaps more than any other area of healthcare, the counselling and psychotherapy profession has reason to be interested in the possibilities afforded by the communications revolution that has swept through society. Telehealth and telemedicine of one sort or another has long excited interest from many areas of health and social care (for example Bennett et al., 1978; Holderegger, 2000; Maheu et al., 2001), but few other types of helping have the opportunity to deliver the crucial elements of the intervention itself by distance means. In the psychological therapies, the possibility of establishing an adequate quality of communication to do that now seems to be becoming realizable.

In recent years, especially, it is noticeable that interest in and use of technology as a means of therapy provision has been rapidly increasing among practitioners from all disciplines and theoretical orientations of the psychological therapies. Media coverage of the idea has increased over recent years, as has the level of demand from clients interested in the possibility of getting the help they would receive from meeting a therapist or counsellor face to face without the difficulties some face in actually having to do so. Some see positive advantages, as many of the chapters in this book testify.

Many of the technologies discussed in this book are not, in themselves, particularly new. Some, such as the telephone, are so well established as to be virtually ubiquitous, at least in most developed countries around the world. Others, such as videoconferencing, are available only to the relatively small minority who have the means to purchase or access the necessary equipment. All, however, are increasingly commonplace and, as the following pages demonstrate, are worthy of our close inspection.

OVERVIEW

The purpose of this book is to provide detailed practical information, backed up with extensive case examples, on the main uses of technology in providing and supporting counselling and psychotherapy. As a result, its range is quite extensive. It covers innovations such as the use of email and Internet chat, the telephone, video links and computerized therapy in the form of software designed for use either with or without the assistance of a trained practitioner. Furthermore, the different contributors look at a variety of aspects of therapy provision. Not only do they offer a variety of perspectives deliberately drawn from their differing theoretical models, they also discuss individual therapy, group work and practitioner supervision whether for trainees and interns or for ongoing supervision for qualified practitioners. It is hoped that the sometimes sharply differing 'voices' in this volume will help to convey the rich breadth of opportunities with which we are now faced and the wealth of contexts around the globe in which these technologies can be applied.

It is, perhaps, a human trait to remark most on that which is new and that which impresses us, rather than turning first to that which is most functional. Throughout this book we have strenuously sought to avoid falling into this commonplace trap. In this setting, technology is not important because it is 'the latest thing' or because it seems to herald a coming age. It is important only inasmuch as it serves a useful function or can be developed until it does so. Thus, the focus here has been resolutely on the practical application of technology and its value. Throughout the book, the needs and interests of practitioners have been given the highest priority, although service managers and potential (or actual) clients of such services are likely to find much that is of interest too.

This book is written in three parts. Part I deals with perhaps the most discussed application of technology in delivering counselling and psychotherapy at present: email and Internet Relay Chat (IRC). Part II considers one of the most well-established technologies and one of its most recent extensions: the telephone and the use of video links or videoconferencing. Part III turns to one of the most controversial, and yet one of the best researched, areas and examines the use of computer programs either to support or even to provide therapy.

Chapter 1, by Kate Anthony, provides a study of how technology has had a role in the profession of counselling and psychotherapy to the present day, from a client and practitioner point of view. This includes the ways in which technology is commonly used in support of mental health services of any kind and provides commentary on the ways in which it is used in the actual delivery of services to clients. Against the background of many years of communications technology innovation, the chapter prefigures some of the discussion in later chapters and examines the full range of devices now available, from traditional telephony, through email and Internet Relay Chat, to fully computerized therapy and videoconferencing. Examples of practice are particularly drawn from email and it is suggested that those practitioners not sufficiently familiar with the new styles of communication, even new kinds of relationships, risk being ill equipped to respond effectively to clients whose lives may be deeply affected by them.

Part I: Email and Internet Relay Chat

In Chapter 2, Peter J. Chechele and Gary Stofle, among the pioneers in offering therapeutic services over the Internet, describe the use of email and Internet Relay Chat ('chat rooms'), suggesting that they are now perhaps the most common use of the Internet in delivering therapy at a distance. Email typically involves a time-lag between responses, that is, a lengthy period can be spent attending to composing and responding to emails at regular, determined intervals. Nonetheless, talking (or, rather, typing) without the immediate responses and interaction that would be expected in other settings is an area that many practitioners find challenging. Internet chat sessions, in contrast, are more immediate, with mere seconds elapsing between a comment typed on one computer appearing on the screen of the other. Chat sessions are thus more similar to the pattern familiar in face-to-face work. Nonetheless, they are usually still restricted to purely textual communication.

One of the most common questions asked of such methods is whether it is possible to create a sufficiently strong therapeutic relationship without being physically present with one another. Some would assert that it is not, despite the long history of epistolary therapy dating from Freud through to the modern day (Bolton et al., in press). In response to such pessimism, while acknowledging that it is not going to suit every client or every practitioner, the authors back up their position that it is possible to create deep and effective therapeutic relationships with examples taken from clients and practitioners who have experienced such relationships for themselves. They are direct about the possible disadvantages of the method and discuss its differences and similarities with face-to-face methods. They note that as well as bringing some new issues for practitioners to consider, such as the increased possibility of either party's words being misconstrued, there are also new techniques and a number of distinct advantages, such as the apparent relative ease with which clients address deeply private concerns and do so more rapidly than would be expected in other contexts.

Chapter 3 looks at a different kind of therapy provision as Yvette Colón and Beth Friedman discuss group therapy over the Internet based on their own work in this field. The chapter opens with a brief discussion of the reasons why some clients may choose to make use of an online group, in favour of other methods. It goes on to describe how the groups operate in practical and therapeutic terms, the options available for practitioners and some of the issues they are likely to encounter. The authors describe the functioning of both synchronous (real-time, chat-type sessions) and asynchronous groups, drawing particularly on examples from the latter method in which a 'bulletin board' system was used in which clients and the practitioner could post messages at any time, from anywhere in the world. This offers the possibility of clients accessing the group 24 hours a day as an ongoing, rolling discussion develops along any number of different threads. The extensive, vivid case material in this chapter clearly documents not only the opportunities and some of the difficulties such groups may encounter but also a number of instances in which the processes that can occur in such groups can parallel the dynamics that would be familiar to any group therapist using more traditional means.

Chapter 4 considers a very different aspect of online mental health provision, and one that has received scant attention in the published literature to date. Michael Fenichel looks at supervision of practitioners, whether they be working by traditional or distance methods, itself provided by email. It is here, perhaps, that the opportunity provided by the Internet to access highly expert practitioners from any location is most obviously of importance. If many parts of the world suffer from a shortage of suitably trained and qualified practitioners in the psychological therapies, there is an even greater lack of those with the additional skills, training and experience to supervise them. Fenichel describes the opportunities and challenges of distance supervision, drawing on a number of sources, including the first experimental supervision groups set up to explore the topic by the International Society for Mental Health Online (ISMHO). Fenichel suggests that some theoretical orientations will be better suited to such methods than others, mediated by the degree to which the processes on which they rely can be replicated through online communication. The chapter provides detailed guidance for those considering setting up or using online supervision as well as considering the challenges and ethical issues involved.

Part II: Telephone and video links

In Chapter 5, Maxine Rosenfield looks at the use of the telephone for counselling and psychotherapy provision. Distinct from telephone helplines, typically used for one-off contacts and advice, guidance or befriending, the telephone as a medium for contracted therapeutic relationships is even now not universally accepted. Perhaps because of the relative dearth of high-quality research, this is still so in some parts of the world, despite the much longer history that telephone counselling and therapy have in comparison with some of the other methods described elsewhere in this book. The arguments in favour of telephone provision overlap with those for other distance methods, with the addition of the different kind of 'intimacy as the counsellor and client speak directly into each other's ears'. Cognitive-behavioural and person-centred or experiential approaches are seen as being especially appropriate for telephone work, although its utility is by no means restricted to them alone. The ability to assess a client's suitability for therapy, to build a sufficiently deep, trusting relationship and offer clients a relatively powerful sense of control and empowerment over the process are all described, with practical guidance for practitioners on the requirements for working in this way. The chapter also briefly discusses the possibilities of group therapy by telephone, the limits of its appropriate use and provides detailed accounts of two cases with a depth of relationship comparable to that which might be expected in similar face-to-face therapy.

Chapter 6, by Susan Simpson, examines the use of video links as a means of providing therapy, with examples drawn particularly from her own cognitive-behavioural practice in the Scottish Highlands and Islands. However, she also surveys the entire range of research on therapy provided by video link from all theoretical therapeutic orientations, setting it in the context of developments in telehealth in general. She concludes that the choice of video link over alternatives, such as telephone or face-to-face contact, appears to be an individual one dependent on personality traits. Technological issues, such as the type and reliability of connection, are briefly discussed before a more detailed examination of the demonstrable effectiveness of such systems and the levels of client satisfaction associated with them. In common with the other methods discussed in the book, it is clear that while some limitations exist, there is a proportion of clients who identify a positive preference for this type of provision. Therapists' satisfaction, which is a much less commonly considered area of psychological research, is reported to be similarly high, with therapists' initial hesitations, typical of most uses of technology in therapy, being overcome as their familiarity with the process increases. Video links, unlike most of the technologies described here, are still relatively new, with most currently available systems still clearly in need of improvements, which are promised with the advent of increased quality and speed of connections provided by the continuing communications revolution. The quality of therapeutic relationship that can be achieved is, the author argues, partially dependent on the quality of the technological connection. We may presume, however, that this is a temporary problem that will be solved as more sophisticated equipment becomes available at increasingly realistic prices. The processes involved in video therapy are discussed and the chapter also contains recommendations for practitioners wth an interest in this area as well as detailed case study material.

Chapter 7 looks at the use of both telephone and video as a means of providing supervision of practitioners. Phillip Armstong and H. Lori Schneiders take, as a particularly testing example, the supervision of trainee practitioners (interns) for their discussion of the use of video links in supervision. An unusual possibility offered by video supervision is that of direct, live observation of a supervisee's sessions with their client from a remote site, with obvious implications for affording accurate, detailed feedback from the supervisor. The authors also consider the possible problems, such as the vulnerability of any data passed over computer networks, balancing the use of live video sessions as a teaching tool. The ability to pick up on subtle nuances of communication by either method is demonstrated by focusing on the more restricted of the two: telephone supervision, which is discussed with examples from the Australian context where the availability of appropriate supervisors and the distances involved in serving some of the most scattered communities on earth make the issues surrounding distance provision especially acute.

Part III: Computerized therapy – stand-alone and practitioner-supported software

In Chapter 8, Kate Cavanagh, Jason S. Zack, David A. Shapiro and Jesse H. Wright provide an account of the history, development and current 'state of the art' of therapy provided through computer programs. Starting from its origins in the late 1960s, they discuss programs designed to imitate a therapist – in one case purely as an exercise in software development – and the later development of systems that demonstrably benefit clients with specific problems such as phobias, depression and anxiety. Some such programs are designed to be used by clients with the assistance of a trained professional while others are intended to be used by clients on their own. The authors point out that despite reactions against the use of computers in this way, in practice there is little reason to think of them as intrinsically different from the self-help books that are generally considered so acceptable as be unremarkable. These more recent products go well beyond self-help, however. Some can personalize what they offer to address the needs of the client more accurately than is feasible with paper publishing or routine, standardized treatment protocols. Language, precise definition of issues and responsiveness to risk and areas of most concern can all be adapted as the client's circumstances dictate. The authors argue, on the basis of at least a proportion of the wealth of research findings presented in this chapter, that in some circumstances computerized therapy may even be better than a human therapist. Controversial though this may be, it is the reports of clients that point to such a possibility. It is perhaps inevitable that such ideas will raise the hackles of some practitioners who would see what they consider to be essential elements of their work, even their humanity, to be under threat from such proposals. However, it is difficult to deny that some clients simply prefer to work on their problems without having to spend hours in the company of a therapist whether therapists like it or not.

In Chapter 9, Kate Cavanagh, David A. Shapiro and Jason S. Zack look at the logistical and ethical issues involved in providing computerized therapy. One of the concerns frequently raised by those looking at the topic for the first time is whether a computer can deal adequately with clients who are emotionally vulnerable or who present a risk to themselves or others. Modern software typically provides adequate safeguards, the authors argue, through such mechanisms as confidential reports to the counsellor, therapist or physician who is responsible for the clients' care. Computer programs do not need to function in a care vacuum in which case management is entirely absent. The ethical requirement to see that, first, we do no harm and, second, that we provide the most beneficial services possible means that some overseeing of their use is always likely to be required. Examples are drawn from several of the most advanced computer programs currently either available or under development, including screen shots and examples of interaction.

In the concluding chapter of the book, we look at some of the themes common to all forms of distance therapy and pose some questions regarding its future. Significant challenges remain before most of the methods discussed in this book can be accepted and we outline a number of issues that must be addressed as a matter of urgency, such as the international regulation of a global profession and the acute need for empirical research for those areas in which it is lacking, before further progress can be made. The conclusion also takes a look to the future technology that may be expected to demand the attention of the counselling and psychotherapy profession, such as virtual reality and sophisticated programming that allows us to create moving, talking therapists that are only embodied by the computer itself.

NOTE ON THE SCOPE OF THE TEXT AND LANGUAGE USED

Appropriate for a subject that, for the first time, provides global access to therapy of all kinds regardless of geographic limitations, this book has been deliberately prepared for an international audience. There are many parts of the world in which the distances between client and practitioner are simply too great to make traditional means of providing mental health care adequate in any cases except those of immense, overwhelming need. It may be that it is simply not economic to provide access to therapy for the whole population. Further, it may be that counselling and psychotherapy services typically have very lengthy waiting lists, implying that any means by which their services can be more efficiently organized are worthy of attention. Extending their 'reach' through distance methods is one such possibility and it is not limited only to isolated communities in marginal areas. Even in areas of dense population, there can be advantages in seeking help from someone outside one's own community or in order to access expertise in a narrow field, available only in another part of the world. We have taken care, therefore, to ensure that the book should be useful and accessible for readers in all parts of the world.

Not only are the contributors deliberately drawn from diverse national backgrounds, reflected in the chapter contents and the case material they provide, the scope and language of the text has been kept as internationally applicable as the great variations in professional contexts around the world allows. We have assumed that the common theme across all psychological therapies is the centrality of the therapeutic relationship (even where that relationship is via, or even with, a technological interface). Broadly speaking, all interventions that might be considered to fall into the category that such an assumption suggests can be considered to be included in this book. At one end of the spectrum we have excluded activities like informal advice-giving and guidance, but have included all formal counselling and therapeutic interventions in both clinical and non-clinical settings, also including formal psychotherapy at the other end of the spectrum. Excluded beyond that, however, has been the realm of medical psychiatry as might be delivered by a hospital psychiatrist, at least in those aspects when it is clearly different from the work of a psychotherapist (for example when prescribing medication).

The title under which psychological therapies are provided varies in itself. The terms 'counselling' and 'psychotherapy' in particular have generally been used interchangeably here in recognition of the large degree of overlap in their use evident in the international research literature (McLeod, 1994, p. 41) but their use should not be assumed to exclude approaches to which the following work might apply.

Those who provide such services go under an even greater variety of titles in different parts of the world. Thus, the terms 'practitioner' or 'therapist' have generally been used, except where a more specific role or function is deliberately indicated. The routes by which practitioners are considered qualified to practice vary perhaps more than any other aspect of the profession from nation to nation and sometimes even between regions within a single country. This is not unproblematic for those wishing to provide or access distance therapy of any regulated type, as discussed further in the concluding chapter. Prior to qualification, intending practitioners may be referred to as students, trainees, interns or a number of other titles. All such designations have been used more or less as synonyms for the purposes of this book.

Even basic phrases like 'mental health' and 'mental health services' are used in different ways in different parts of the world. Here, they are used to indicate any provision intended to be of benefit to the mental wellbeing of the client group, regardless of the severity of their condition. Thus, we would include acute psychiatric services but would not restrict such terms to them, also including the whole panoply of possibilities from acute care through therapeutic counselling and clinical psychology to more informal counselling and even some support-type services.

Despite occasional references to counselling and psychotherapy in relation to medical contexts, those who seek help from such interventions have generally been referred to as 'clients', rather than as 'patients', except where a distinct medical relationship is being referred to. Thus, a medical doctor (physician) may see a 'patient', but the same 'patient' would generally be referred to in this book as a 'client' when seeing a therapist or counsellor. Grafanaki (1997) cites Hoyt (1979) and Hill and Corbett (1993) who suggest that the term 'patient' 'relates more narrowly to a medical model, whereas "client" seems consistent with the counselling focus on strengths over pathology' (p. 4). We have followed this convention on the same ground of not excluding a humanistic emphasis and common practice in the field of counselling and psychotherapy generally.

Finally, it should also be noted that although this book contains extensive material gleaned from case studies and excerpts of therapy in progress, all the material has been thoroughly anonymized and all identifying features have been altered to maintain confidentiality.

SUMMARY

New technology in counselling and psychotherapy is an exciting and controversial area of innovation. No area of the profession should proceed ahead of the evidence base and for that reason we are neither seeking to endorse any of the methods nor dissuade colleagues from their use and exploration. This book seeks to stimulate discussion about the possibilities of using technology in therapy and give an account, including case material, of what can currently be done and how it can be done well. While the use of technology in therapy is increasing, it is too early to say that this is clearly an unmitigatedly good thing, despite the reasons for optimism offered by the various contributors to this book. Practitioners are out there doing it and clients are out there seeking it. This is a situation that the profession must address and that practitioners interested in the area should not ignore.

REFERENCES

Adams, D. (1999) *How to Stop Worrying and Learn to Love the Internet.* Available at http://www.douglasadams.com/dna/19990901-00-a.html.

Bennett, A.M., Rappaport, W.H., Skinner, F.L., National Center for Health Services Research and Mitre Corporation Metrek Division (1978) *Telehealth Handbook: A Guide to Telecommunications Technology for Rural Health Care*, US Dept of Health,

Education, and Welfare, Public Health Service, National Center for Health Services Research: available from NCHSR Publications and Information Branch, Hyattsville, MD.

Bolton, G., Howlett, S., Lago, C. and Wright, J. (in press) *Writing Cures: An Introductory Handbook of Writing in Counselling and Therapy*. Brunner Routledge, London.

Cogswell, J.F. and Estavan, D.P. (1965) *Explorations in Computer-assisted Counseling*, System Development Corp., Santa Monica, CA.

Cogswell, J.F., United States Office of Education and System Development Corporation (1967) *Exploratory Study of Information-processing Procedures and Computer-based Technology in Vocational Counseling; Final Report*, System Development Corp., Santa Monica, CA.

Grafanaki, S. (1997) Client and Counsellor Experiences of Therapeutic Interaction During Moments of Congruence and Incongruence: Analysis of Significant Events in Counselling/Psychotherapy, Unpublished PhD thesis. Keele University, Keele.

Hill, C. and Corbett, M. (1993) 'A perspective on the history of process and outcome research in counselling psychology', *Journal of Counselling Psychology*, **40**(1): 3–24.

Holderegger, J. (2000) *Bridging the Gap: Telehealth in Rural America*, Wyoming Development Disabilities Division, Wyoming.

Hoyt, M. (1979) 'Patient' or 'client': what's the name?, *Psychotherapy: Theory Research and Practice*, **16**(1): 46–7.

Maheu, M.M., Whitten, P. and Allen, A. (2001) *E-Health, telehealth, and Telemedicine: A Guide to Start-up and Success*, Jossey-Bass, San Francisco.

McLeod, J. (1994) 'The research agenda for counselling', *Counselling, Journal of the British Association for Counselling*, **5**(1): 41–3.

Rogers, C.R. (1942) 'The use of electronically recorded interviews in improving psychotherapeutic techniques', *American Journal of Orthopsychiatry*, (12): 429–34.

Weizenbaum, J. (1966) 'ELIZA – A computer program for the study of natural language communication between man and machine', Proceedings of Conference of Association for Computer Machinery. New York.

1 The use and role of technology in counselling and psychotherapy

KATE ANTHONY

INTRODUCTION

The use of technology in counselling and psychotherapy is changing the face of the profession, with practitioners either being challenged and excited by the new opportunities it presents or feeling sceptical, overwhelmed or even frightened by an unwelcome intrusion into traditional methods of providing mental health services. There are valid reasons for both points of view. Embracing technology as the way forward without pause for thought on how safe or effective it is or how it can be done ethically – even whether it should be done at all – is at best naive and at worst dangerous for clients and practitioners alike. But ignoring technology, as it becomes more and more a part of our daily lives, suggests ignoring, firstly, the client's *choice* of receiving mental health assistance in this way and, secondly, the suggested *possibilities*, such as improved efficiency, that technology may afford the profession as a whole. The use and role of technology in counselling and psychotherapy must be recognized as worthy of our attention and exploration.

This chapter therefore addresses clients' and practitioners' exploration of technology in two ways. It assesses the impact that technology has had on our working lives and looks at its use in creating therapeutic relationships. What this chapter does not attempt to address is the myriad ways that technology allows adaptation of hardware and software for better access to such communication facilities for clients and practitioners who have specific disabilities. Examples include those who have hearing problems and rely on sign language to be able to communicate using technology such as videoconferencing, email or a 'minicom', and those with sight impairment or blindness being able to use adapted text-to-speech software. Websites can be available in large type, enlarged text documents and colour contrasted text, and computer mice exist that allow the user to 'feel' website hyperlinks through vibration to the hand. There is also the range of adaptations that can be made for people experiencing paralysis or debilitating disease. Such technology can itself be used in ways that facilitate access to therapy, but it is beyond the remit of this chapter to address whether or not this should mean that technological ways of delivering therapy are more appropriate.

Technology has increasingly impacted on the way that mental health practitioners conduct their day-to-day business. Communication on a professional level between two parties can be faster, more efficient and more convenient from an administrative point of view (including the use of general office equipment such as answerphones, fax machines and pagers as well as email). The use of mobile phones allows for geographical access in most areas of the world. The use of 'personal digital assistants' (PDAs, also referred to as palmtops or handhelds), laptops and notebooks allow for scheduling and communication away from the office or consulting room, as well as access to information via the Internet, wireless application protocol (WAP) or general packet radio service (GPRS), among other telecommunication systems. Use of a computer for creating and storing notes, information and databases is becoming increasingly acceptable. Accessing and conducting research and training is convenient and simple, alongside using offline resources such as informational CD-ROM software.

Using technology in what has traditionally been considered 'the *talking* therapies' creates a need to be careful as to how we define each aspect of these new ways of working. It has sometimes seemed the case that the profession saw the flashing lights of the computer or heard the whirr of the video equipment and could not see past them as to what was actually happening between client and practitioner. What is clear, once we get over our fears of using technology in counselling and psychotherapy, is that we are examining the same phenomenon that has happened in all types of therapy in the last 100 years: the fact that *communication* between two parties is the key to finding mental wellbeing in the face of our life circumstance. We are still human beings interacting with each other to try to cope with problems; we are just communicating using a different set of tools. This factor holds for all the technological methods discussed in this book, even where the client is using a piece of software designed by practitioners – the skills and experience of professionals are still being communicated to clients.

Interaction with another person or persons using technology has four distinct parameters:

1. *without* any visual or audio experience of each other (as when communication is exclusively by text)
2. a *purely* audio experience of each other (telephone)
3. with visual and audio experience *taking place at a distance* (video-conference)
4. with *remote* visual and audio experience taking place (where one party uses software developed by the other – as discussed in Part III of this volume).

The interaction may happen in real time (synchronous communication such as using the telephone or Internet Relay Chat (IRC)) or may be subject to time delays (asynchronous communication using email or CD-ROM software). But the impact of technology on the profession is also under scrutiny from the perspective of the practitioner's interaction with the hardware and software itself. We are learning to make better and better use of technology and bend it more to our will – once successful we stop calling it technology, as quoted from Douglas Adams in the Introduction.

MEDIA-LED AND DATA-STORAGE TECHNOLOGY

We may see the invention of the telegraph by the Chappe brothers (see for example Holzmann and Pehrson, 1994) and print by Gutenberg (see Eisenstein, 1983) as the first examples of technology being used to impart information to the masses. Cinematic news bulletins and propaganda such as the Pathé Newsreel became available in 1909. The development of radio and television allowed for distribution of expert opinion and mental health programming in many forms. Programming in the last 50 years has ranged from serious documentaries on specific issues all the way through to entertainment 'reality TV' posing as psychological experiment. While the former can have a useful function in conveying ethical and considered discussion, the latter has meant that practitioners and governing bodies have needed to consider the impact of such shows on potential client participants and their families (Hodson, 2002). Increasingly bold news footage of events around the world such as war, terrorism and natural disaster has also meant that their impact reaches a far wider community than those directly involved. Satellite and cable technologies convey almost instantaneous pictures and sound, and interactive television allows for personal choice of how and when information is received.

Video recording also allows us to choose what to watch and when. The use of handheld amateur video cameras and CCTV systems allows an extra dimension in being able to access distressing footage, from the Kennedy Zapruder film through to the more recent 'last moments' of victims of child abduction and murder. All these uses of what we no longer think of as advanced technology to receive potentially psychologically potent information can impact on clients and practitioners within mental health globally, and are entirely media-led.

Video and audio technologies also allow us to tape live sessions with clients for research, supervisory and training purposes. Examination of the process of a session after the event can reveal hidden dynamics (for example Rogers, 1942; Slack, 1985) and allow close attention to detail such as interpretation of body language and tone of voice as well as examination of issues such as questions of ethical behaviour. Informed consent from the client on the creation and use of such tapes is paramount (even with such familiar uses of technology, safety is an issue and we may do well to note that it is now routine, to the point where it does not even raise an eyebrow), but the wealth of data that can be gathered from the study of what happens between client and practitioner within the therapeutic hour is invaluable to our understanding of how and why counselling and psychotherapy has a successful or unsuccessful outcome (see Orlinsky and Howard, 1986, for example). These methods of taping sessions are now mirrored in more recent technology where the Internet (email or IRC) session is or can be *automatically* stored and processed by both practitioner and client, as discussed later.

TELEPHONE TECHNOLOGY AND PORTABLE TELECOMMUNICATION DEVICES

In 1875, Alexander Graham Bell was granted the UK patent that allowed two signals to be sent at the same time in an attempt to improve the telegraph. A year later, his 1876 US patent described his invention as

> the method of, and apparatus for, transmitting vocal or other sounds telegraphically by causing electrical undulations, similar in form to the

vibrations of the air accompanying the said vocal or other sounds. (see http://www.invent.org)

Telephone communication, and more recently other telephonic software development such as Internet telephony (VoIP – voice over Internet protocol), is based on what we now refer to as 'landline' technology. The telephone is usually synchronous (although asynchronous communication is possible with messages on answerphones, fax machines or pagers) and allows us to cross global barriers and time zones. Other barriers are broken down, including restricted access to face-to-face communication through disability, remote geographical locations or life circumstances that prevent the client or practitioner leaving the home. Eliminating the need to travel for therapy can be valuable for many clients. Picking up a telephone and calling a mental health service provider can in itself be an empowering move for clients. They have the ability to receive therapy in the comfort and familiarity of their own location, avoiding the sometimes intimidating experience of a practitioner's waiting and consulting rooms. This may allow clients to be more relaxed and therefore able to open up more easily. They have the choice of terminating the session at any time simply by hanging up, which is perhaps generally easier than leaving the room in a face-to-face relationship. There is often also a heightened sense of bravery from the client, who is likely to self-reveal more quickly when not having to face someone more directly. For this reason, telephone therapy is often seen as tending to involve shorter contracts than face-to-face therapies (Rosenfield, 1997). A discussion of the use of the telephone specifically for counselling and psychotherapy can be found in Chapter 5.

Another application of telephone technology is computerized telephone systems, where clients use the buttons on their telephone pad to assess their condition (for instance, for obsessive compulsive disorder through the BT *STEPS* program, Marks et al., 1998) and design and implement their own self-help programme, being guided and assisted through it by the automated telephone system. Discomfort at exposure or level of ritualistic behaviour can be monitored, with voice backup being arranged through recorded messages to request immediate, live input from a practitioner. Progress reports can be sent to the client and the practitioner, detailing development and successes and therefore affirming to the client their success.

In addition to these considerations, the impact of telephone technology reaches much wider than just that of landline communication with clients or between practitioners. While mobile phones may arguably be considered unsuitable for actual therapeutic work due to the unreliability of their signal and possible privacy issues, their widespread use for verbal communication allows people to access each other in most parts of the world and has almost reached saturation point in many countries, to the point where we no longer consider them to be unusual. Alongside the use of mobile phones for talking to each other came the unexpected popularity of short message services (SMS), or 'texting'. This often heavily abbreviated way of messaging another person or group is already second nature to generations that have grown up with mobile phones being the norm rather than the exception. Clearly there are implications for practitioners who are unfamiliar with their client's use of technology to form, develop and maintain personal relationships (Anthony, 2001). The ability of mobile technology to carry emotional impact is demonstrated in negative form by examples such as its use to bully peers, which has led to suicides in a number of countries (see http://www.successunlimited.co.uk/related/mobile.htm), and also as a tool for stalkers (von Heussen, 2000). Recent developments for mobile technology include the ability to send pictures, which may be used for similar negative activities rather than their intended positive social functions, both of which will be of importance in understanding client issues in the future.

COMPUTER TECHNOLOGY AND STAND-ALONE SOFTWARE

PDAs and laptops/notebooks are also becoming mainstream for practitioners in the mental health profession to simplify appointment administration and exchange of data between practitioners. As telecommunication methods improve, access to the Internet, email and WAP sites allows for a truly coherent technological interface from which to organize a mental health service or operation. The ability to carry large files of data and information in a piece of hardware no bigger than an adult hand or a ring binder is a welcome benefit for many people. Synchronization between desktop, laptop and handheld data is increasingly simple, with networks allowing remote access to information held at a main base. There is a long list of computer software applications of use to counsellors and psychotherapists in developing and maintaining their individual or organizational service to clients. They include basic software such as using a word-processing package (for correspondence, note-taking or writing up research, for example); a database package (for client or staff information); and using spreadsheets to manage and study data for both quantitative and qualitative research.

In the late 1960s, Joseph Weizenbaum developed a piece of software he named ELIZA (see online version available at http://www-ai.ijs.si/eliza/eliza.html) to emulate a stereotypical Rogerian therapist, using a basic natural language engine that recognized keywords and could reply in hard-coded stored sentences. This was done to show how a computer could convincingly imitate an actual person, in this case a therapist, and pass what is known as the Turing Test (see http://cogsci.ucsd.edu/~asaygin/tt/ttest.html). This stated that for a computer to be considered intelligent, it needed to convince a person that he or she was talking to another human being. ELIZA's communication was very basic and yet, to the horror of its inventor, its users developed meaningful relationships with the program, sharing sensitive information and considering the exchange to be therapeutic, even after becoming aware that they were interacting with a piece of software (O'Dell and Dickson, 1984). Use of ELIZA can be frustrating because of its limitations. Yet the technological advances made since its invention now allow us to make use of the ultimate distancing of therapists from clients – complete therapist absence.

Indeed, the use of CD-ROM technology (and, more recently, DVDs) is an important development in the use of personal computers in mental health services. They can be used to develop and distribute software for computer-aided psychological diagnosis, cognitive rehabilitation, delivery of cognitive-behavioural therapy (CBT) and biofeedback to help manage and reduce anxiety and depression. Removing the therapist from the therapeutic input is not as dramatic as it sounds. The input from mental health professionals to software development for the delivery of more structured and directive therapies gives wide access to assistance where resources are limited. The software can be designed to guide the client through the process of face-to-face CBT – problem-focusing, setting goals, monitoring emotion and activating event reactions, allowing for setbacks and training in coping techniques. All this data can be digitally stored and printed for the client to work with between the sessions. Clients can also sometimes see carefully prepared examples of other people's problems and coping strategies and identify with them, creating role models. In addition, the software can incorporate natural language engines to react to client's keywords, particularly with regard to recognizing potential dangers to the client (such as 'I want to kill myself'). Questionnaires, based on traditional methods of mental health assessment, such as the Beck Depression Inventory scoring system, are very easily implemented. It is the case that, whether practitioners like it or not, use of computers rather than personal interaction is preferred by some clients. For a more extensive discussion of CD-ROM technology, see Chapters 8 and 9.

COMPUTER TECHNOLOGY AND THE WORLD WIDE WEB

In 1991, the work on developing the World Wide Web (WWW) by Tim Berners-Lee and others at the Conseil Européenne pour la Recherche Nucléaire (CERN) was completed. The CERN team created the protocol based on hypertext that makes it possible to connect content on the Web using hyperlinks. In his short history of the Web, Berners-Lee (2000) describes the team's dream of the future of society through use of the Internet:

> The dream behind the Web is of a common information space in which we communicate by sharing information. Its universality is essential: the fact that a hypertext link can point to anything, be it personal, local or global, be it draft or highly polished. There was a second part of the dream, too, dependent on the Web being so generally used that it became a realistic mirror (or in fact the primary embodiment) of the ways in which we work and play and socialize ... we could then use computers to help us analyse, make sense of what we are doing, where we individually fit in, and how we can better work together.

What has developed since then is a wealth of information in millions of personal and professional websites, requiring selective and objective evaluation of content and value. With this in mind, mental health practitioners have available to them, at the click of a mouse and the use of a search engine, extensive information (or misinformation) and detail on any given subject of interest. The potential ease of continuing professional development in this way has opened up new avenues of understanding more about client issues and circumstances from a global cultural perspective, particularly in developing countries. It is worth noting that these areas of the globe often have sufficiently well-developed technology, and so the potential for working with diverse cultures where access was previously limited has issues of its own. Finding information has been further simplified by the use of 'natural language engines', whereby programs are able to recognize not only the words provided by the user but also the context of the words, how they relate to each other and the implication of those words when placed in that particular order (Anthony and Lawson, 2002). Thus, you can ask a search engine a question in everyday language instead of having to use keywords or knowing how to use some of the more complex forms of searching such as Boolean expressions.

The increasing ease with which websites can be created means that practitioners can develop their own information sites for clients seeking assistance for their problems. These can contain internal and external hyperlinks to provide a simple route for clients to gain further help. Many provide self-help questionnaires which will guide the client, as through a flow chart, to find advice and guidance about what treatment may be appropriate for them before being able to make an informed decision as to what road to take in receiving treatment. Initial assessment may also be conducted via a questionnaire on the Internet and use *automatic* processing of data to screen potential clients who would not be considered suitable for online therapy and who are then routed to more appropriate sources of help (see www.interapy.com for example; Lange et al., 2000). In addition, registering with search engines means that the practitioner's ability to reach a client base is extended globally, allowing much cheaper advertising than paper-based notices and listings.

Client use of websites can also mean that they maintain public Web logs ('blogs') of their own material, usually uploaded on a daily basis, giving accounts of their thoughts, emotions, life events and so on. While these may be useful to the practitioner in gaining a rounded picture of a client's state of mind (and in supervision, as discussed in Chapter 4), they may also include commentary on the therapy in progress and the therapist. Documented cases (Fenichel et al., 2002) include demonstrations of the client's ability to post public messages on a practitioner's professional website becoming problematic as the client became jealous of the practitioner's public interaction. It is the practitioner's responsibility to consider what stipulations need to be in place within the therapeutic contract to establish what would be appropriate for public dissemination.

Listservs and newsgroups are bulletin board systems where messages can be posted and other people can respond. These played a major role in the history of the development of online mental health provision as groups like alt.support.depression became early places where sufferers could share thoughts and feelings (Zack, 2002, personal communication). Many thousands of mental health e-groups (where practitioners can subscribe to an open email discussion forum in individual or digest form) also exist for the dissemination of opinion and information, a process that is examined further in Chapter 3. More recent development of software to facilitate online therapeutic relationships between client and practitioner includes email held on secure servers on the Internet which are password-protected, bypassing the need to send emails to each other at all and therefore limiting the possibility of third-party intervention (see Screenshot 1.1). This also gives a website-based transcript of the therapeutic process which is easier to follow than a trail of sent mail held on home PCs and paper versions (http://www.forumnexus.com, for example).

In addition to accessing information, gathering literature and carrying out research through websites, conducting both qualitative and quantitative research over the Internet has developed into an at least sometimes appropriate way of gathering mental health data. Mann and Stewart (2000) identify four initiatives in Internet research:

1. standardized interviews in the form of email and Web page-based surveys
2. non-standardized forms of online one-to-one interviewing
3. observation of virtual communities
4. the collection of personal documents online.

Quantitative surveys can be submitted electronically over the Internet anonymously and can include open-ended questions for qualitative analysis. Online interviewing can be carried out without the identity of the respondent being revealed, facilitating more open responses on topics of a personal nature. This author's original research on the nature of the therapeutic relationship in online counselling and psychotherapy (Anthony, 2000) was conducted via the second method, with all communication and information-giving taking place via website, email and IRC. A further advantage of this method is the loss of the need for having the material transcribed from tapes or notes, since all the data was gathered and retained electronically, saving valuable research time.

Specialist (post-qualification) training taking place online is also an important new technological application for practitioners. This developing phenomenon allows global interaction between the trainees, accessible 24 hours a day with the use of software hosted on the

Screenshot 1.1 Example of practitioner's view of secure client correspondence mail boxes

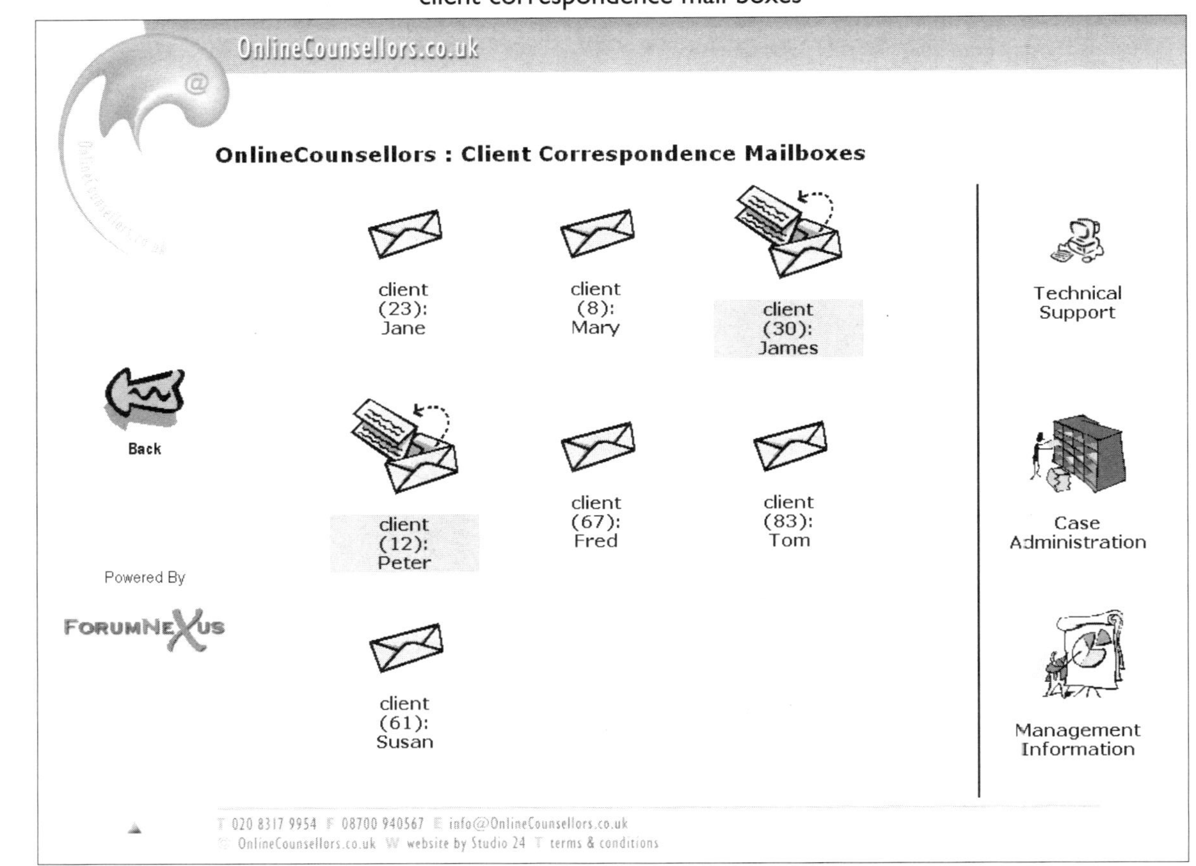

Screenshot 1.2 Example of online training programme (therapy using chat rooms module)

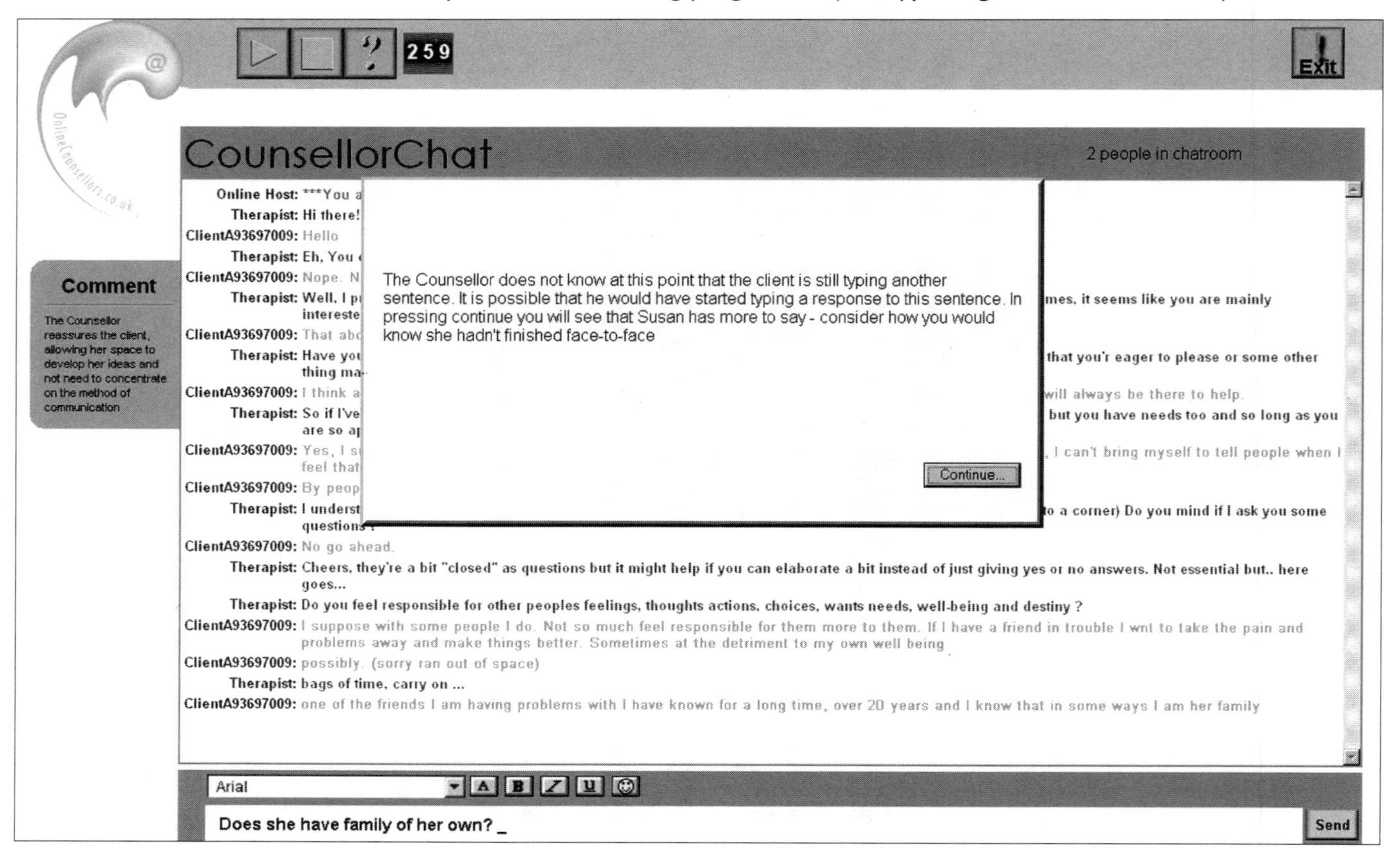

Internet for specific subject areas and discussion boards. Coursework can be held on the server and accessed at anytime, creating a paperless route to learning material. Learning or tutorial groups can meet in chat rooms or via videoconference from anywhere in the world, allowing cultural issues within mental health work to be better considered and experienced. In addition, access to mental health 'experts' may be better facilitated, including the possibility that where the usually heavy workload of professors and other academics prevents them being able to give opinions via interview to students and researchers, a short polite email request is often responded to. Password access to live therapy sessions is possible, and a number of structured programs that lead the practitioner through the trail of quality and evidence-based research posted on the Web are now available in many areas of mental health provision (see Screenshot 1.2, www.onlinecounsellors.co.uk, for example).

COMPUTER TECHNOLOGY AND EMAIL

With the invention of the Internet came perhaps one of the most useful communication tools of the century – the creation of an electronic version of mail, universally known as email. This fast and convenient method of sending messages to individuals and groups became widespread globally during the 1990s. In a fascinating exploration of the changing language of email, Danet (2001) identifies the two traditional templates used for email – the memo template for business use, and the letter-writing template for personal use. Both exhibit obvious and differing social norms, including the content of the subject line, the use of emoticons such as the smiley :-) or :o) (best understood by inclining one's head slightly to the left), abbreviations such as u instead of you, and the use of lower case initial letters for informality and upper case FOR SHOUTING. These norms became known under the collective description of 'netiquette', a truncation of Internet etiquette. Further examples of netiquette can be found in Chapter 2.

The first recorded fee-based use of encrypted email for long-term online therapy dates back to 1995 to practitioners such as Dr David Sommers, Dr Leonard Holmes, Dr Gary Bresee and Ed Needham (see www.cyberchoices.org or www.ismho.org). Guidelines and suggested principles started to appear with the formation of the International Society for Mental Health Online (ISMHO) in 1997, and other organizations such as the National Board for Certified Counselors (NBCC,

1997), the American Counseling Association (ACA, 1999) and the British Association for Counselling and Psychotherapy (BACP) (Lago et al., 1999). Perhaps the most detailed such guidance to date was produced by the BACP (Goss et al., 2001).

Practitioners working with email need to be aware of how the lack of sensory input when working with text means that the words have to be supplemented with explanation, including adding the elements of netiquette that make up, in part, for lost sensory clues. Indeed, it is easy to misunderstand the typed (often described as 'stark') communication, and it is here that the practitioner needs to be aware of their client's likely tendency to interpret the meaning of any number of phrases in various ways. For instance, consider the following client text:

> I mean, I wouldn't want anyone to think I was being stupid, that would be awful, wouldn't it?

The therapist may, with the best intentions, respond thus:

> They may think you were being stupid. They may not think you were being stupid. We could concentrate on whether you think you are being stupid and why it would be awful.

The client, already feeling anxious and defensive at being thought stupid, may read the text in the way it was meant, but could also read it with a selective eye for emphasis, which reinforces the original thought pattern:

> They may think **you were being stupid**. They may not think **you were being stupid**. We could concentrate on whether you think **you are being stupid** and why **it would be awful**.

The use of careful conversational rephrasing is important to create a recognizable 'tone' as would happen in a face-to-face or telephone environment, where the voice could be heard. This further clarifies the intended meaning of the therapist, frequently important in email or IRC communication in which misunderstandings are relatively easy to create, albeit inadvertently:

> Well, they may think you were being stupid. They may not. I can't read their minds. What I'm interested in is whether you think you are, and why it would be awful. I don't think you are, but that's neither here nor there.

Enhancing the text will further minimize chances of misinterpretation and give the client an opportunity to 'hear' the therapist's voice and be able to create a picture of what facial expression, such as a quizzical look or a smile they may have had were they face to face:

> Well, * they * may think you were 'being stupid'. They * may * not. I can't read their minds <unfortunately>. What I'm interested in is whether you think you are, and why it would be 'awful' <really??>. I don't think you are, but that's neither here nor there :o)

Once an email has been sent, a verbatim record of the entire process of the communication exists at several points in cyberspace. Indeed, this aspect of working online seems to be a concern to practitioners who are wary of using technology for therapy. The potential for threats to confidentiality is one concern; the chance for their therapeutic intervention to be held up for scrutiny is another. One of the added benefits of working with email, for both practitioner and client, is the ability to revisit and revise the text to better reflect what is meant before it is sent. Misunderstandings through the text not being revised to achieve better understanding can happen very easily, but may also provide opportunities to examine projection and possible transference reactions, which can be effectively achieved (Suler, 1998) with skill and care within a solid therapeutic relationship.

This ability to revise and correct text can also result in losing what may, hypothetically, provide some interesting therapeutic material. For example, a client whose behaviour towards her husband was increasingly manipulative and underhand, could consistently type 'slly' instead of 'silly'. On a first reading of the client's email, the therapist could read the word 'sly', giving a quite different interpretation of the phrase 'Maybe I was being silly'. These 'Freudian typos' (Anthony, 2000) can give an added depth of insight into the client's maladaptive behaviour, but the ability to revisit the text (not to mention the use of a spellchecker) can mean that such rich interpretative material is lost.

In addition to Danet's memo template for business use and the letter-writing template for personal use, perhaps this new form of communication demands a third definition of an email template – that of the therapeutic email (see Screenshot 1.3). At once both a personal communication because of the nature of its content and purpose and a professional transaction between practitioner and client, we struggle to find a comfortable description of it. Experience of using

Screenshot 1.3 Example of therapeutic email (practitioner response)

From Geraldine Wilkes - Message (HTML) - US-ASCII

File Edit View Insert Format Tools Actions Help

Reply Reply to All Forward

From:
To: ClientA93697009@aol.com
Subject: From Geraldine Wilkes
Sent: Tue 17/12/02 17:16

Hello John,

Thank you for taking the time to tell me about yourself. I imagine it wasn't an easy task to review your situation and set it all down for a complete stranger to see. You seem concerned that perhaps you have given me too much background and I wonder whether the reason for this is that you feel that in some way you have taken up too much of my time. I want to let you know that I was very interested and moved by your story and that the detail has helped me to see you and understand your circumstances a little more clearly. I hope that during the time we work together that will increasingly be so.

So why was I moved by your story? I was struck by how alone in the world you are. You gave me the information "divorced, no kids, family all dead" in what seemed a very unemotional, detached kind of way and yet I note that you describe yourself as "emotional". My guess is that these losses have had a big impact on you and from what you say have left you feeling isolated and lonely. I wonder if you have any other feelings associated with losing or (in the case of children) not having had these significant people? When I try to picture what lies behind the stark statement my sense is that maybe you are feeling hurt too. Is that your sense of it?

You sign yourself as "John the idiot". I felt angry when I read that. John, I want you to know that this may be the way you choose to see yourself but I'm not buying into it! You are a human being and as such have my respect. And in any case, nobody who can write about themselves with such clarity and impact can be regarded as an idiot. If it is an attempt to get me to feel sorry for you then I'm afraid that won't work either though I do feel sad that you seem to be so self-critical. You don't seem to be giving yourself any credit for attaining a university degree or running your own business and these are just the things you have told me about. Are there any more? You seem very perceptive and self-aware and have developed some coping strategies to make up for what you lack in self-confidence. You also seem very aware of the dissonance between how things are in your life and how you would like them to be. At the moment this seems to be the cause of much unhappiness but I'm wondering if you could see this incongruence in a different light - perhaps as a motivator for change?

One thing I hear loud and clear running through your message is your anger. While it seems that sometimes it is directed outwards (maybe towards the women that you perceive took advantage of you) it often seems to be turned inward (is this the real reason for the "idiot" self descriptor?). What I don't know is whether you are able to express this anger. I wonder how anger was dealt with in your family when you were growing up. In my experience it is very common for it to have been somehow unacceptable or wrong to get angry and as I expect you know anger that doesn't find an outlet and is turned inwards can often lead to anxiety and depression. Maybe you would like to spend some time thinking about this and, if you feel it applies to you, perhaps finding safe ways of expressing your anger now. Is there some unfinished business, something you would like to have said to your parents, ex-wife or ex-girlfriends that you could find a way of articulating now?
You seem to regard yourself as a passenger rather than in the driving seat of your life with "being miserable" as something that has been ordained for you. I wonder where you learnt that you don't have a right to get your own needs met or that you can have so little impact or influence over what happens to you. How did you learn that you have very little entitlement to another's affection or care and attention? I am thinking here of the way you stopped having therapy after such a short time because you thought you were wasting your therapist's time and you were not prepared to spend the money on yourself.

I can see that the depression you have retreated into is a miserably cold and lonely place. And yet I guess there is a certain safety there of not having to risk the pain of being hurt by another. I am really sorry that relationships have not worked out and that you need to protect yourself from further hurt in this way. Is this why you have only taken "stabs" at making relationships in more recent years?

You have asked a number of questions. If you are hoping that I will provide the answers then I need to tell you that I do not regard myself as the expert on what goes on in your head. I believe that somewhere may be deep down you know what is best for you and you have the ability to find the answers that are right for you. I hope this doesn't feel that I am abandoning you to work it all out by yourself. I can and will accompany you and support you as we explore these issues together if that is what you want.

I hope this has been of some help or at least got you thinking.
I look forward to hearing from you next week.
Best wishes
Geraldine

email for conducting a therapeutic transaction suggests the need for practitioners to have three separate email addresses or accounts to maintain boundaries between personal, business and therapeutic communications. Within 30 minutes of being logged into one standard, general-use email account, a practitioner can receive several client emails, which can be interspersed with personal emails that usually need only a quick response, business emails that need to be put aside for more studied consideration and several 'spam' or junk emails that simply need to be deleted. By maintaining separate accounts, not only is the risk of deleting important mail diminished, but also the time allocated to client material can be better maintained and dedicated, as would be the case in a face-to-face session. Most Internet service providers (ISPs) offer the facility to have multiple email addresses as standard, with each email address having its own password protection.

The benefits of email therapy for clients who are unable to access face-to-face therapy may seem obvious. Powell (1998) found that the flexibility of the service provision and the ability to access services ranked highest as the strongest advantages; it could be for geographical reasons, for reasons of disability, or due to life circumstances such as being in an abusive relationship where seeking treatment outside the home would be difficult. Suler (1999) also comments on the ability to 'test the water' before going into therapy, with online work serving as an initial step towards face-to-face therapy in a safe, possibly anonymous way. However, as stated with the other forms of distance therapy discussed in this book, it may also be the case that some clients prefer not to be with their therapist in a physical space at all and, in addition, some may even be better equipped to work with text rather than face to face. This is generally seen as the result of the client not having to 'look someone in the eye' when revealing sensitive material. There is also the matter of clients being able to be as comfortable as possible in their own space and working at their own pace, not necessarily constrained by the 'therapeutic hour'. Emails can be composed, stored, revisited, reworked and sent at any time. Without the need to pay attention to the usual social norms of dress and timing, more time may be spent on the message's meaning (Walther and Burgoon, 1992). The concept of making an appointment is made irrelevant in most email work, although, in practice, boundary issues mean that practitioners are well advised to see that the therapeutic contract states that emails will be replied to within a specific period of (for example) a week. These aspects of online work for therapy are discussed further in the next chapter.

COMPUTER TECHNOLOGY AND INTERNET RELAY CHAT (IRC)

At around the same time as email was being applied to mental health encounters for the first time, ongoing therapeutic relationships as well as meetings between practitioners were also being created via IRC. Working in chat rooms to hold a conversation holds different challenges. Where email offers the opportunity for quite rational articulation of material, IRC is much more spontaneous and occurs in real time, with both parties being online at the same time. Individual style varies from person to person and it is up to both parties to find a level at which they are happy with the length of each exchange before waiting (usually a few seconds as the written message travels across the Internet) for the reply to appear, although with improved Internet connections this will reduce dramatically (Screenshot 1.4). However, most people, even those competent in typing, cannot communicate as quickly as they could with the spoken word. For clients and practitioners alike, there is also the reappearance of having to make an appointment (perhaps across time zones). Chat rooms may need to be adapted to have visual characteristics to make them seem less stark than is often the case with email (by use of colours, fonts and graphics, for example). Again, a verbatim transcript of the conversation can be logged on a PC and printed off and each practitioner will have their own uses of these when it comes to making notes for client work. Chat room therapy is also further discussed in Chapter 2.

As with email, the practitioner needs a knowledge and understanding of how the Internet community exists and functions. Communities spring up and disband very quickly within group chat rooms, as do the relationships within them. Therapists operating in any sphere are increasingly likely to come across clients who have experience of these and will need to have at least a basic awareness of their potential. For example, chat rooms can sometimes offer opportunities for 'cybersex' (defined by Hamman, 1996, as computer-mediated interactive masturbation in real time and computer-mediated telling of interactive sexual stories (in real time) with the intent of arousal). 'Cyberinfidelity' or 'cyberrelationships', where clients may develop intense personal intimate relationships with people across the world that they have no actual physical knowledge of, can sometimes have devastating effects on marriages and other 'real-life' (known in Internet communities as RL) relationships. The ability to maintain day-to-day functional mental health can also be affected, for example if an addic-

Screenshot 1.4 Example of therapeutic chat room session

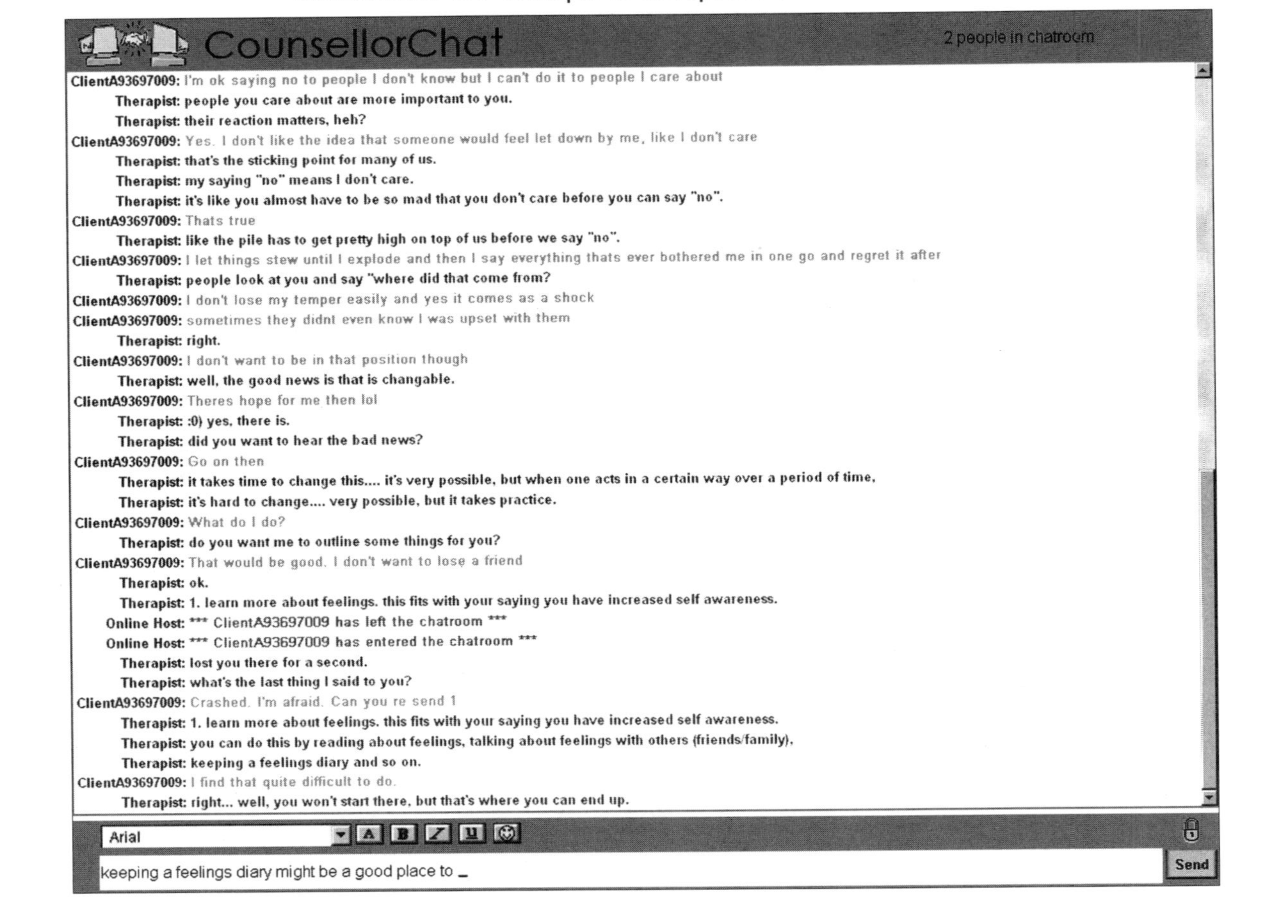

tion to these relationships surfaces (see Young et al., 1999; Maheu, 2000 for examples). The psychological effects of virtual rape in cyberspace also need attention for a rounded view of just what it is possible to experience, perhaps especially when in an advanced pictorial online community, where it is possible to 'pin' a character (or 'avatar') down and use software manipulation to make them perform and speak obscenities (see Dibbell, 1993).

Identity management in cyberspace is the subject of intense speculation from a social and sexual point of view in chat room facilities of various kinds. Comprehensive studies can be found in Suler (2000) and Sempsey (2000). Online, it is far easier than in face-to-face encounters to alter who you appear to be or what parts of your real self you allow to show and it can be taken to much greater degrees, including outright impersonation. It is possible, and can sometimes be unproblematic, for clients to present themselves under the guise of any identity they wish. It is sometimes useful to examine the motivations behind the choice of log-in name or avatar (graphical representation of identity, sometimes referred to as an 'av'). While identity-switching may hide potentially important issues, the ability to explore and express aspects of the self that may otherwise remain less prominent has been seen as a positive benefit of distance therapy provision, especially when through text-based media. This may well be an issue for research into group therapy online (as discussed in Chapter 3), especially where there is the ability to act out negative personality traits in a disruptive manner while being safely at a computer terminal. Conversely, it takes a good group facilitator to draw out the 'lurker' within the group who is content to sit back and watch the discussions unfold. The ability to 'hide' behind an avatar, log-in name or handle are reasons why we can suspect that text therapy is likely to have a role even when better video or other telecommunication links are available.

COMPUTER TECHNOLOGY AND WEBCAMS

The first use of a 'video-frame grabber' on a pre-Web intranet (1991) was to monitor when a coffee pot of the Computer Science Department of Cambridge University was full and establish whether it was worth making the long journey from other departments to refill one's cup. In 1993 they experimented with regularly updating the image over the Internet and so it became the first live webcam (see Stafford-

Fraser, 1995). Since then, many one-way webcams are used for viewing various live video-streams, mostly for viewing landmarks or, more infamously, accessing live pornography.

A more relevant use of webcams and video equipment in mental health is two-way communication. For practitioners, this means that face-to-face meetings between individuals and groups of mental health teams are possible globally, transcending the usual constraints of distance and access. In addition, face-to-face supervisory relationships can also be facilitated, as discussed in Chapter 7. Researchers can now arrange for international face-to-face interviewing without leaving their homes, with implications for fund and budget allocation being improved both personally and from governing bodies.

Video links have been used for trials of conducting both physical and psychological therapies over the Internet (for example Ross, 2000; Greenlaw et al., 2002), with the benefits – and limitations – universal to distance therapies of access to services. The difference when considering using video links in comparison with other means of practitioner communication over the Internet is that it may be considered to be a limited way in which clients and practitioners can return to working face to face yet with better facilitation of access. Videoconferencing standards, both from a point of view of equipment and connection speeds, will improve and seem set to become mainstream over the next few years, with costs plummeting to make it accessible to practitioners and clients wishing to access therapy from their homes or a provided video suite. This is discussed further in Chapter 6.

CONCLUSION

With access to technology literally at our fingertips, there is a wealth of information, communication and interaction available to us in increasingly accessible avenues. It remains to be confirmed whether these current technological interfaces are a positive step for the profession of counselling and psychotherapy, but their role is defined and already present, even ahead of a solid research evidence base for us to refer to. Indeed, it seems that some clients' choice of technology over face-to-face communication, and the services offered by practitioners and organizations to meet that demand, has necessitated a rush towards a frantic discussion about their possibilities. As more and more sophisticated technology develops, the profession may always be seen to be trying to catch up with it to find an ethical, practical and theoretical stance from which to support it. It is fortunate that the increased efficiency and improved global communication that technology gives us *also* increases our resources and opportunity to develop the research base from which we can achieve this support. This may well be the real challenge for the profession.

REFERENCES

American Counselling Association (1999) 'Ethical Standards for Internet Online Counseling'. Available at http://www.counseling.org/resources/internet.html.

Anthony, K. (2000) 'Counselling in cyberspace'. *Counselling Journal*, **11**(10): 625–7.

Anthony, K. (2001) 'Online Relationships and Cyberinfidelity'. *Counselling Journal*, **12**(9): 38–9. Also available online at www.kateanthony.co.uk/research.html.

Anthony, K. and Lawson, M. (2002) 'The Use of Innovative and Virtual Environment Technology for Counselling and Psychotherapy'. Available online at www.kateanthony.co.uk/research.html [accessed 10 Feb 2003].

Berners-Lee, T. (2000) 'The World Wide Web: A Very Short Personal History'. Available online at www.w3.org/People/Berners-Lee/ShortHistory.html [accessed 29 July 2002].

Danet, B. (2001) *Cyberplay*. Berg, Oxford.

Dibbell, J. (1993) 'A Rape in Cyberspace'. *The Village Voice*, December 21, 36–42. Available online at www.apocolypse.org/pub/u/lpb/muddex/vv.html [accessed 6 June 2002].

Eisenstein, E. (1983) *The Printing Revolution in Early Modern Europe*. Cambridge University Press, Cambridge.

Fenichel, M., Suler, J., Barak, A., Zelvin, E., Jones, G., Munro, K., Meunier, V. and Walker-Schmucker, W. (2002) 'Myths and Realities of Online Clinical Work'. Available at www.fenichel.com/myths [accessed 2 July 2002].

Goss, S., Anthony, K., Jamieson, A. and Palmer, S. (2001) *Guidelines for Online Counselling and Psychotherapy*. BACP, Rugby.

Greenlaw R., Wessel, D., Katevas, N., Andritsos, F., Memos, D., Prentza, A. and Delprato, U. (2002) 'PARREHA. Assistive Technology for Parkinson's Rehabilitation'. Available online at www.eng.cam.ac.uk/cwuaat/02/14.pdf [accessed 6 June 2002].

Hamman, R. (1996) 'Cyborgasms: Cybersex Amongst Multiple Selves and Cyborgs in the Narrow Bandwidth Space of America Online Chat Rooms'. Available online at www.socio.demon.co.uk/Cyborgasms.html [accessed 2 October 2001].

von Heussen, E. (2000) The Law and 'Social Problems': 'The Case of Britain's Protection from Harassment Act 1997'. *Web Journal of Current Legal Issues* Blackstone Press Ltd. http://webjcli.ncl.ac.uk/2000/issue1/vonheussen1.html. [accessed 30 July 2002].

Hodson, P. (2002) 'BACP Ethical Guidelines for Production Companies', Press release issued 27 July.

Holzmann, G.J. and Pehrson, B. (1994) 'The Early History of Data Networks'. Available online at http://vvv.it.kth.se/docs/early_net/ [accessed 13 August 2002].

Lago, C., Baughan, R., Copinger-Binns, P., et al. (1999) *Counselling Online Opportunities and Risks in Counselling Clients via the Internet*. BACP, Rugby.

Lange, A., Schrieken, B., Van de Ven, J-P., Bredeweg, B., Emmelkamp, P.M.G., Kolk, J. van der, Lydsdottir, L., Massaro, M. and Reuvers, A. (2000) '"Interapy": The Effects of a Short

Protocolled Treatment of Post-Traumatic Stress and Pathological Grief Through the Internet.' *Behavioural and Cognitive Psychotherapy*, (28): 175–92.

Maheu, M. (2000) 'Women's Internet Behavior: Providing Psychotherapy Offline and Online for Cyber-infidelity'. Available online at http://cybercounsel.uncg.edu/featurearticles/artofweek031500.htm [accessed 7 March 2001].

Mann, C. and Stewart, F. (2000) *Internet Communication and Qualitative Research: A Handbook for Researching Online*. Sage, London.

Marks, I.M., Shaw, S. and Parkin, R. (1998) 'Computer-aided Treatments of Mental Health Problems'. *Clinical Psychology: Science and Practice*, (5): 151–70.

NBCC (1997) 'NBCC WebCounseling Standards', *Counselling Today Online* at http://www.counseling.org/ctonline/sr598/nbcc_standards.htm [accessed Feb. 1999].

O'Dell, J.W. and Dickson, J. (1984) 'ELIZA as a Therapeutic Tool'. *Journal of Clinical Psychology*, (40): 942–5.

Orlinsky, D.E. and Howard, K.I. (1986) 'Process and Outcome in Psychotherapy' in S.L. Garfield and A.E. Bergin (eds) *Handbook of Psychotherapy and Behaviour Change*, 3rd edn, John Wiley, New York.

Powell, T. (1998) 'Online Counseling: A Profile and Descriptive Analysis'. Available online at http://www.netpsych.com/Powell.htm [accessed 7 May 2001].

Rogers, C.R. (1942) The use of electronically recorded interviews in improving psychotherapeutic techniques. *American Journal of Orthopsychiatry*, (12): 429–34.

Rosenfield, M. (1997) *Counselling by Telephone*. Sage, London.

Ross, C. (2000) TBC 'Counseling by Videoconference'. Paper presented at the 5th International Conference on Client-Centred and Experiential Psychotherapy, Chicago, June.

Sempsey, J. (2000) 'The Therapeutic Potentials of Text-Based Virtual Reality'. Available online at www.brandeis.edu/pubs/jove/HTML/v3/sempsey.html [accessed 5 March 2002].

Slack, S. (1985) 'Reflections on a workshop with Carl Rogers'. *Journal of Humanistic Psychology*, (25): 35–42.

Stafford-Fraser, Q. (1995) available online at http://www.cl.cam.ac.uk/coffee/qsf/coffee.html [accessed 12 August 2002].

Suler, J. (1998) 'Mom, Dad, Computer: Transference Reactions to Computers'. Available online at www.rider.edu/users/suler/psycyber/comptransf.html [accessed 29 March 2000].

Suler, J. (1999) 'Psychotherapy in Cyberspace: A 5-Dimension Model of Online and Computer Mediated Psychotherapy'. Available online at www.rider.edu/users/suler/psycyber/therapy.html [accessed 29 August 2001].

Suler, J. (2000) 'Identity Management in Cyberspace'. Available online at www.rider.edu/users/suler/psycyber/identitymanage.html [accessed 21 September 2001].

Walther, J.B. and Burgoon, J.K. (1992) Relational communication in computer-mediated interaction. *Human Communication Research*, (19): 50–88.

Young, K., O'Mara, J. and Buchanan, J. (1999) 'Cybersex and Infidelity Online: Implications for Evaluation and Treatment'. Poster presented at the 107th annual meeting of the American Psychological Association, 21 August.

Part I

Email and Internet Relay Chat

2 Individual therapy online via email and Internet Relay Chat

PETER J. CHECHELE AND GARY STOFLE

This chapter discusses the emerging field of individual online counselling and psychotherapy conducted through text using email exchanges and real-time Internet Relay Chat (IRC).

In November 1997, the first author tentatively hung his 'cybershingle' and opened the website www.cybertherapy.com for business, unsure of what would be encountered. The sense back then was that there was a vast potential for bringing improved mental health to numerous individuals who might never have set foot in a therapist's office. Now nearly five years and several hundred clients later, the potential of this unique and exciting method for the delivery of mental health services seems even more convincing. Throughout the chapter, direct quotes from actual online clients are used to better illustrate the concepts being examined. All the names and identifying information have been changed so as to protect the confidentiality of clients, and the text appears as it did in the original, including misspellings and typos.

People have always found comfort and connection in writing letters and keeping journals (see http://www.journaltherapy. com/links/index. html for a brief history). Emailing and the Internet are a modern form of this. Quietly gathering ones feelings into written communications allows individuals the opportunity to bind their anxiety and gain some distance and perspective from their problems. Email therapy may be considered a sort of 'talking journal' where the individual can share his or her thoughts with an empathic other.

The challenge that faces mental health professionals who are interested in offering these types of service is how to translate the skills used in face-to-face therapy to a text-only medium. Clinicians working from many models are quite dependent on verbal and non-verbal cues and may have more difficulty assessing clients through text alone. In contrast, therapists who envision their role as more of a teacher, motivator and collaborator rather than healer may struggle less with this issue.

Although it has tremendous potential, online therapy may not be suitable for everyone. Certain prerequisites improve the likelihood that an individual might benefit from these services. The experience of the first author indicates that the best candidates for treatment are those individuals whose difficulties can most probably be remedied through a combination of insight, education and practised behavioural interventions. Clients who are suicidal or have a serious mental disorder such as schizophrenia or bipolar or have a borderline personality disorder are more difficult to treat and may need more intensive services than can be offered online.

Since the work is conducted through text alone, it is useful if the client is a proficient typist (although this could be more applicable to chat room work) and is comfortable communicating through text. The client should have a basic comfort level with technology and be reasonably knowledgeable about using his or her computer. Ideally he or she will have some tolerance for computer glitches and can endure periods of silence between communications. Although difficult to assess, the client should be emotionally strong enough and psychologically tough enough to be able to work through miscommunications and projections. If at any point in the treatment the client appears to be deteriorating, he or she should be encouraged to seek out more intensive services in their geographic area. A client should be willing to work within the agreed-upon frame and contractual agreement. And last but not least, in many if not most situations, the client must have the ability to pay for services. At this time, third-party reimbursement does not cover these services in the US, and most health insurance schemes do not cover these services. However, it is possible that this will vary increasingly from country to country in the light of what cost reductions can be verified as a result of their use.

It is appropriate for clinicians working in this modality to be adept typists and have more than a beginner's grasp of the Internet and in communicating with others online. They too have to be able to tolerate periods of silence and be able to work through the miscommunications that inevitably happen. The clinician should know how to use appropriate encryption software and privacy technology to protect client confidentiality and safeguard records. Also important is having the skills to be able to extrapolate through text the heart of the client's concerns. It is important for clinicians to remain grounded but flexible in their approach and do everything possible to maintain the agreed-upon contract. Finally it is imperative that clinicians develop competence in handling acting-out behaviour and intensity of emotion as expressed in text communications. The following excerpt came from a client who experienced this author's reply as an empathic break and reacted strongly with the words below. Being able to work through these types of break can be challenging:

> your words didn't address the intense emotions from those emails. its like I came up to you and just got hit by a truck and my guts were hanging out and blood everywhere and you basically said ... hope you are happy, have a great day.

SIMILARITIES AND DIFFERENCES

Significant differences exist between face-to-face and email therapy. Although both seek to help individuals to gain a better understanding of themselves and improve their coping skills, the way this is achieved is noticeably different. Face-to-face therapists typically attempt to accomplish this through weekly 50-minute sessions, while online therapists strive to achieve it through weekly email exchanges. With the former, clinician and client agree to a certain location and time of day when it fits within both their schedules. Email exchange communications occur at any time of day from any computer with an Internet connection. 'Asynchronous communication' is the term used to describe this type of alternating email exchange:

> One of the reasons I like on-line counseling is that I feel I am able to express my feelings better in writing than in person. I'm finding that one of the down sides is that it's hard to maintain the normal ebb and flow of a conversation.

Therapy is not dependent on when the therapist is available but when the client feels the need to share, and then the therapist is able to choose the best time to respond appropriately:

> To me, therapy has been war, at times. But this experience, so far, is different. Why? Mainly because I don't have to shut up when our once-a-week hour is over.

With email therapy being devoid of non-verbal cues, clinicians must ask a lot more follow-up questions to clarify uncertainties in the communications. Over time they may begin developing an eye for 'reading between the lines' of these exchanges. On the other hand, the absence of these cues may free up clients to discuss difficult issues that previously they were too self-conscious to bring up in the therapist's office.

The virtual anonymity of email exchanges allows clients to focus on their issues without being distracted or overwhelmed by the sensory stimuli confronting them in a face-to-face session. Clients appear more eager to address their issues and feel more in control and safer behind the keyboard of their home computer:

> I just find it easier to write than to talk. Sometimes I forget words when I talk and I end up stuttering, or I lose my train of thought. If I'm writing, I can look back at what I wrote and pick up what I was going to say.

In the experience of the authors and with brief therapy generally, online therapy seldom lasts longer than a month or two. Dependency on the clinician is kept to a minimum, with clients being encouraged to take active control of their lives. Clients know that if their life situation should change in the future they can always check back in for additional support. Therapists with brief orientations have the advantage of motivating clients to get involved quickly in treatment while instilling hope that change can occur in a relatively short period of time. Normalizing the fact that mental health is always in flux and that periodic check-ins are to be expected, especially during periods of undue stress or transition, provides additional object constancy for the client:

> I like the thought that support is out there when I might need it. My plan will be to contact you as I feel the need. I think that just the fact that I know I can contact someone if I need to is soothing to me.

Additionally, where face-to-face therapy traditionally happens in the therapist's office, online therapy takes place in the client's own space, both physically and emotionally. It also has more of an aspect of crisis intervention because the emails contact them much more in the centre of their crisis; clients have the ability to contact the therapist while they are in the midst of their emotional turmoil. Having a clinician respond in a timely manner provides more immediate access to interventions, which may reduce the duration of the clients' feelings of vulnerability, as opposed to waiting for the weekly therapy session. The client below emailed her therapist on a Sunday night after a particularly difficult weekend. The reply on Monday morning found her still wrestling with her emotions:

> Thank you for your email. As I told you last night this weekend was an emotional roller coaster for me. This morning I cried when I read your reply, I think it's finally sinking in … anyway I guess I should pull myself together now and get to work, I'll email you later with more.

Even clinicians who are adept at taking thorough notes during or after face-to-face sessions will ultimately miss important details of these sessions. In contrast, doing therapy through text communication provides both the clinician and client with the entire notes verbatim. Clients find it particularly beneficial to be able to muse over this text at their leisure and often find deeper meaning with repeated readings:

> I was re-reading some of my earlier emails to you. When I wrote these and read your responses I felt you didn't care for some reason. But now that I re-read them I do see that you did care a lot. I guess I couldn't feel anything but the pain back then. IT seemed so deep and anything you said to me seemed superficial at the point in time, but now it doesn't seem that way.

This advantage is also shared by the clinician who can quietly sift through and reflect on these exchanges undistracted by the presence of the client. When clients return to therapy months or years later to discuss current stress in their life, the clinician need only reread a few pages of email exchanges to reorient themselves to the client's life situation. It should be noted that if records should ever be subpoenaed these would be verbatim records of the therapy.

Additional reasons why clients might prefer to 'be seen' online include social phobia, dealing with an intensely shameful issue or because they fear judgement due to race, sexual orientation or socioeconomic status, for example. They may have a physical disability that makes access difficult or they are housebound. It may be that they have never been to a psychotherapist before and are curious. If there are no therapists in their locale, online therapy could be of great benefit. Or they may simply wish their therapy to be more anonymous:

> I am a therapist myself. My reason for exploring online therapy has several factors: First of all, I live in New Hampshire which is comparable probably to living in Mill Valley or someplace out there where all the professionals know each other, and those you don't know you I probably wouldn't want to go see.

Whatever reasons there are for seeking out these services, the numbers of individuals entering online therapy continues to grow. More research and training must be done to better address these needs and prepare clinicians for the challenges ahead.

PRACTICALITIES AND CONTRACTING

Online clinicians need to address several 'housekeeping' issues before therapy can commence. The first and most important issue is informed consent. Many online clinicians have an informed consent/disclaimer form on their websites which the client must agree to before proceeding with treatment. There are many examples of informed consent issues that the client will need to know. These could be the differences between email therapy and face-to-face therapy and mention of where the therapy is understood as being conducted. Limits of confidentiality and options for securing emails and files are also important. In addition, there are several legal and ethical considerations such as the practitioner's chosen policy for handling grievances, the termination policy and mention of mandated reporting laws. Finally, unless a specialist service, it should emphasize that these services are not intended for suicidal clients. Once the client has read and agreed to these issues, further questions will need to be addressed, such as determining how the communications will take place (that is, via email or chat), whether to contract for one session or a longer contract, and what the fees will be and how they will be collected. How to handle technology problems and what to do in cases of emergency is also important. Clients may have questions about these and other issues and the time to address these concerns is at the outset of therapy.

An explanation of the fee structure, if appropriate, should be clearly stated on the website. Some clinicians continue to bill their services based on an hourly rate, which is determined by the amount of time it takes them to read and respond to the client's emails. Others contract for a set fee for a single email, a package of emails or a monthly retainer. Chat sessions are typically billed on either a per-minute basis or per session. Payment options should also be readily identifiable on the website.

Keeping communications private and records confidential is very important. Although 100 per cent confidentiality is not a reality in either the face-to-face world or online, there are security measures such as using a secure socket layer (SSL) for the submission of sensitive material and financial information. Widely available encryption software can be used to safeguard email and protect client records. Developing an organized filing system for keeping track of active and archived clients' records can be achieved using one of several email programs. Regularly backing up these records on a removable tape or zip drive is imperative as hard drives are prone to crashing and all client documentation could be permanently lost.

Some therapists have an online intake form to gather clinical information for screening out individuals who might need a more intensive level of services. Others screen potential clients by offering them a free initial email inquiry, after which they follow up with specific questions to expand understanding and address concerns that may affect the treatment. The Internet is an international phenomenon and the 'virtual office' is open to anyone who has access to a suitable computer and an Internet connection. It is important to take into consideration the language capabilities, cultural backgrounds and sociopolitical climate of potential clients before deciding on whether you are knowledgeable enough to treat them. Organizations with responsibility for overseeing or regulating the quality of counselling and the psychotherapies (such as state licensing boards or professional associations) may have issues with therapists 'seeing' someone outside their geographic boundaries. Furthermore, therapists should not treat someone online that they would not feel competent treating in their office.

Despite the best efforts to select appropriate clients, occasionally either a 'good match' does not develop or the client may need more intensive services, and practitioners need to be prepared to offer alternative referrals or resources. Online clinicians differ widely in the amount of demographic information they gather at the outset of therapy. It is good practice to have at least the client's telephone number and an alternative email address, for when technology glitches happen or clients are in crisis, and, in turn, the clinician should also provide clients with several options for contacting them. The therapist's website should also have links to crisis resources such as The Samaritans (http://www.samaritans.org.uk) or www.metanoia.org/suicide/. Providing links to additional self-help resources is also good practice. Internet resources have the ability to include international users as the norm and need not exclude people because of distance in a crisis.

With email therapy it is important to be timely and consistent in responses. Many online therapists recommend a response time of between 24 and 48 hours. Clients find it comforting to know when they might anticipate a reply – if too much time lapses between exchanges the anxiety of the client and/or the clinician may negatively affect the work.

THE WORK

There are numerous methods for developing a therapeutic relationship through text. Mood and personality can be suggested through the use of various fonts, the choice of colour, THE USE OF CAPS, as well as the actual structure of the sentences and words. Some therapists like to respond to the client's words line by line as if the client were making the same statements in their office (see case example). Replying in this way ensures that most things the client has said will be addressed and the response will sound more conversational and directly connected to the client's words. The use of colour or different fonts makes it easy to differentiate the therapist's words from the client's. In face-to-face sessions clinicians are often presented with multiple issues from the client. Clients who are anxious, not well organized or who may be trying to avoid emotionally charged material might present the clinician with a flurry of verbiage. Decisions are constantly being made in face-to-face therapy as to which thread to pursue and which to ignore. In contrast, email therapy allows the client the opportunity to talk uninterrupted about as much or as little as they desire. The clinician then has the opportunity to explore multiple issues, having complete access to the client's flow of thoughts. Eventually the clinician will want to focus these discussions towards one or two central themes, but need not totally disregard lesser subjects. Typically, after response to much of what the client has sent, the comments are reread and edited for clarity and emphasis. With experience, there is less need for rewriting and one's initial response can be trusted to a greater degree. An introductory paragraph prepares clients for the reply by summarizing the central issues, with thoughts for them to consider. Finally, some practitioners like to give considered thought to the subject line of the email, and may use inspirational quotes or affirmations that relate to the central theme being discussed (see Cook, 1993). For example, for a client who had been raised by a critical parent and now was having difficulty trusting her own ability to raise her children, the subject line might contain:

> Subject: 'I am not afraid of storms, for I am learning how to sail my ship' – Louisa May Alcott

One way to join with the client is by using his or her own words in the reply:

> From the age of 7, I remember feeling very depressed and unhappy with myself. I felt silly, foolish, lazy, fat, ugly. Lately I feel like **I'm a fraud**; and

> **very childish** because I don't just buckle down and get on with my life. [authors' emphasis]
>
> *That little girl must have felt so sad and misunderstood. Her belief in the future was compromised by these feelings. And now that this child is all grown up it's difficult for her to believe in herself.*

If done well, this empathic mirroring can make the client feel heard and encourage additional disclosure.

With its absence of visual cues, therapeutic work over the Internet is ripe for projection. Frequently at the beginning of online therapy the therapist is idealized. It may only be after a number of email exchanges that the client begins to perceive the therapist in a more realistic manner. When the projection breaks down there is always the potential for strong feelings in the client. Working through these feelings is often a part of the early work. If clients are to be in charge of their own healing, it is important that they do not give up their power.

As clients begin to assume more responsibility for the direction their life will take, the therapist can begin challenging them. Pacing is important, especially given the tendency for clients to open up quickly online. The clinician needs to be cautious in the early exchanges so as not to overburden clients with more than they can assimilate. In email therapy, indications of the client's ego strength or readiness to address charged material can be surmised by paying special attention to what issues and in what manner the client responds. If the client chooses to ignore a therapeutic intervention or move the email thread in a new direction, this might indicate that the clinician may need to rethink their timing or the manner in which this material is presented.

After exchanging a few emails with a client, a rhythm starts to become established and the next response can usually be anticipated. During periods of extended 'silence', the clinician should respond with an email asking how the client is doing. With experience, one may also begin to notice when clients' writing seems to deviate from their normal discourse. They may start writing less, the content may change or they may become disorganized or tangential in their replies. Asking probing questions on the nature of these variations often reveals new areas of psychosocial stress in the client's life. Along with being able to offer support around this new stress, addressing such changes also lets the client know that you are paying close attention to them.

One of the advantages of working online is having the Internet at your fingertips. Whether it is used for looking up information to help in treatment planning, suggesting books or articles for your client or consulting with a colleague, the Internet is alive with potential resources, although it should be noted that misinformation is also widely available. However, with good sourcing, much information may be imparted to the client.

TERMINATION

Termination happens when clients feel that they no longer need clinical support, they are not finding benefit or they move from working online to working with a clinician face to face. By contracting with clients for a relatively short time, say a month at a time, there is a built-in review process at the end of each contract period. If, at the end of these 30-day periods, the client and therapist feel that additional contracts are warranted, then a new contract period is agreed upon. If at any time during one of these contract periods the clinician gets the sense that the client's functioning is greatly improved, he or she may suggest a hiatus to allow the client the opportunity to test his or her newly developing skill set. Clients know in advance that they can always check back in at a later date should additional support be needed. It is important for clinicians to be honest in their assessment of how much change clients can realistically expect over the length of a contract. Intermittent time-limited therapy allows for clients to check back in during particular periods of stress, such as going to see their MD or GP when they develop a new illness.

THERAPY USING CHAT ROOMS/IRC

Clients occasionally prefer one mode of communicating over another and which they choose and why can be significant. Beginning with email affords both clients and practitioners more time to gather history on the clients and formulate hypotheses regarding their presenting problems. Many clients prefer to use just one modality for communicating whereas others will move from one to another depending on many variables. To generalize, clients who prefer chat usually want a more intimate real-time connection that can mean they feel closer to the therapist. Often they have an issue that has some sort of urgency to it and would like immediate feedback concerning it. There seems to be a spectrum of closeness from email to chat to telephone and finally to face-to-face counselling. Clients will sometimes enter therapy using email and move up the continuum, eventually arriving in a therapist's office for face-to-face therapy.

Sometimes it is nice to be able to sit back and reflect on the words, thoughts and feelings of your client and at other times it is nice to have the revealing and spontaneous interactions that chat rooms provide. Although clients can self-censor themselves in either format, typically you get more unconscious and/or 'real' material during a live chat session. Therapy in a chat room presents some different problems and issues from therapy using email. Both the therapist and the client need to be online at the same time and have their computers working properly to be able to access the chat room. It is still essential to have offline contact information for clients so that they can be contacted in the event of a hardware/software failure at the scheduled chat time, whereas with email, the therapist is under less pressure, in the sense that another computer can usually be found in a reasonable length of time to compose a reply. Chat therapy, because it in some ways approximates to face-to-face therapy, can be used for either short-term or long-term work with clients. As with any therapy, it is essential to develop with the client a contract for a number of sessions and then provide the client with a structure and framework for the chat sessions. As with email therapy, similar assessment procedures apply.

Once in the chat room, the therapist needs to be able to develop a sense of the client's communication timing and tempo and has to be able to increase his or her concentration precisely at a time when he or she has the freedom not to concentrate. If the therapist becomes distracted, it can limit the ability to tune into the client's tempo and respond accordingly (Stofle, 2002). Evaluating the timing of a client's response can be important and also provides diagnostic information. The more difficult the issue, usually the more time it will take for the client to respond. The therapist needs to be respectful of the client's need to take time to respond, while being aware that the client may have been distracted by something completely unrelated. A client's timing of responses needs to be evaluated constantly in the chat session.

As in face-to-face work, the chat room therapist needs to be able to think on his or her feet in order to be able to respond in the moment to what the client is presenting. However, in a chat room the therapist has a short amount of time to rehearse and try out various responses to the client. With email therapy, the therapist can have the luxury of a long period of time to think and contemplate a response. In this regard, both chat and email offer a distinct advantage over face-to-face therapy where one cannot take back what one has said. Of course, whatever the modality of treatment, the therapist works with the client through a continuing dialogue to discover the true meaning of the client's words.

In both email and chat therapy, the written word is the most important thing. The therapist needs to be fluent in the language (often briefly and quickly – see case example) to get the salient point across to the client. The words can be quite stark – without any adornments – and this starkness can let the therapist respond directly, honestly and powerfully to the client. Chat therapy offers an opportunity for clarification of communication in the moment, which can prevent misunderstandings. It is important for the therapist to seek clarification if there is any doubt regarding the client's message. The therapist needs to ensure that the client also has appropriate communication skills.

EXAMPLE OF EMAIL THERAPY

Diana, a 39-year-old divorcee, contacted an email therapy service in 1999 when she was struggling with feelings of unworthiness and longing for acceptance. She stated that she was having difficulty finding a therapist near her location in Montana, and felt unable to focus on or complete projects. Her parents had divorced when she was young and she felt constantly rejected by her mother, resulting in a fear of rejection in the present. She stated that she functioned well but wanted to realize her full potential. In the contracting email, the therapist noted that she seemed the perfect candidate for online work. This email was a normal one – a letter from the therapist to her. Below is a hybrid of the email exchange that followed. The plain text is her response to the therapist's final contracting email, but the bold text (usually in a different colour in the original) was interspersed with that response. (Note: In earlier emails, this client was assessed as an appropriate candidate for online therapy. She was not suicidal; she demonstrated the ability for self-reflection and was functioning fairly well in her current life situation.)

Subj:The important thing is somehow to begin – Henry Moore

Thanks for the response...**That's why I'm here.** Well, it's nice to be the perfect candidate for something...ha...What do you think would be the most effective way...email, chat or phone? I am stretching it financially to do this at all and as much as I like chat, I'm afraid I would be too 'aware' of the 'time' passing, and I don't want to be thinking about money during our sessions...**I agree starting with a series of emails sounds like a good way to begin...you can use them as**

you feel the need. Chat does seem to fly bye [sic] and I wouldn't want to see you spend that time thinking about dollars and not dynamics

Five sessions contracted to start sounds reasonable. It should give both of us a chance to see if we can move forward with this. I expect, however, that I'll be doing this the rest of my life. I really want one person to work with, as I 'think' it will be easier for the therapist to understand what is going on with me at different times if they know me better…and that takes time. **Yes it does take time to get to know someone and although taking care of ones mental health is a life long concern, I believe we can make some good progress in the days ahead.**

I 'think' chat is effective because you don't have time to think…you would be getting first thoughts rather than contemplated thoughts…am I wrong? **There are pros and cons to each way of communicating…the immediacy of chat does seem to limit the amount of self censorship, whereas email allows you time to organise your thoughts so as to better communicate your ideas. They're both good for each of these reasons.** And, because I flip-flop so much, I wonder if what I present in my email might be irrelevant by the time I get an answer. **If it is immediacy you are after then chat would meet that need, however keep in mind that any chat session would have to be scheduled and an issue you are grappling with today might be better dealt with through email than waiting for a chat session we have scheduled for two days from now**

Well, I feel like I'm just chattering and not making any sense… **Your making perfect sense, it's normal to be a little self conscious and scattered at the outset of therapy…relax we'll eventually bring this all together.** so, let me know what you think.

It would be helpful if in your next email you included the following so as to help me get to know you better and for us to begin narrowing the focus of our work:

- **More details about your individual and family history.**
- **What do you perceive as your strengths and who do you have in your life for support (emotional and/or financial)?**
- **What do you want the outcome of our work together to be? How would you know when that has happened?**

- **An alternative email address and or phone number to reach you.**

That will be a good start and if you have some questions as to my background, feel free to ask. Just so you'll know, I often use this type of format when responding to emails...that is I usually type along side your comments with my own thoughts and then you can do the same with a different colour or font...It sort of creates a living document that evolves as we proceed. When it starts getting unruly, usually around the end of a series we can start a new exchange.

Looking forward to your reply...

Peter

After Diana's reply to this email, she began corresponding with the therapist quite frequently. Diana seemed thirsty for positive affirmations. At times she emailed so frequently that the therapist was forced to cut and paste several of her emails into one reply. The therapist was being deluged by her insatiable appetite for a mirroring self-object. Self-objects are so vital to a person's functioning that they are experienced as if they were part of the person. When self-objects don't respond to our needs, tremendous anxiety and frustration can result (Seruya, 1997, p. 16). It didn't take long before Diana began perceiving the therapist's delay in responding as proof that she was not worthy of acceptance. She began firing off emails asking the therapist if she was too much for him and whether he still wanted to work with her. The therapist felt the work was proceeding quite nicely, and her testing behaviour could be confronted. The following shows how the therapist was able to cut and paste Diana's emails and respond to her points as a whole session:

Subj: Everyone believes very easily whatever they fear or desire. FONTAINE, JEAN DE LA (1621–95, Poet)

Dear Diana,

I'm having a hard time keeping up with you...I really do want us to approach this relationship at a slow to moderate pace. In my experience it works best that way. It gives both of us a chance to think about what we are saying before continuing on to the next thought. I realise that sometimes it is difficult to wait 'out there' in cyberspace wondering what is happening on this end and I will try and keep to my word as to when you can

expect to hear back from me. What I'd like to do now is to cut and paste bits and pieces of your last few emails and talk a little about them. Then I'll finish with a general summary about what I think has been going on between us this past week. Here goes:

I still expect you to bail out of this. (after reading your email, I cried – that's a rare occurrence).

Diana this expectation or fear of rejection is a very powerful dark force in your life...I think you go back and forth between wanting very badly to feel close to someone and to find acceptance yet at the same time get scared that you might get abandoned. So you continually test those that attempt to get close to you. The sad thing about this way of interacting is that you may end up scaring off people who might otherwise enjoy your company. And on top of this when those individuals leave it strengthens your belief that no one could really love you in the first place. In reality what is happening is that they don't see through your fear...they don't see that lost lonely child that was ignored and rejected by her mother.

I don't want to become dependent upon you, or anyone else, for my emotional well being. That in itself is frightening.

Yes it can be quite frightening putting your trust in others. Whereas an infant has no control over the disposition of their caregivers...an adult certainly does. It's a scary proposition making yourself vulnerable to possible rejection. You were hurt bad once...maybe it's easier to protect yourself from future hurts by keeping others at a distance?

My friend can call the therapist any time he needs or wants. He doesn't, other than regular sessions, unless he is in a crisis and without an appointment. Maybe that fact makes me feel less valuable.

Although I can't be available any time you need to talk, I'm wondering what I can do to help you appreciate your true value?

You are doing a wonderful job. I am so happy that I found you. If our relationship ends today (and God only knows, I hope it doesn't), I will be grateful for what I have reaped from it.

Along with feeling warm and cuddly inside when I hear this, I also am struck by the extremes in your perceptions of me. It

seems I'm all good or all bad. And I'm really somewhere in between.

I don't understand how or what I'm doing to change the things that need to be changed, but I 'feel' changes occurring...I just hope it's not just a temporary state of mind.

Funny thing about therapy but it seems to work this way... things sometimes change and yet you have a hard time figuring out how it happened. One thing to keep in mind however...you never really 'get rid of the ghosts in your closet' but you can learn to recognise them and keep them under control.

Peter, I feel like my next email from you is going to say that you don't think we should work together. Am I right?

The simple answer is no but after what we've already shared, can you tell me why you'd be thinking these thoughts? Diana you have so much anxiety related to being rejected (just had a thought about one of your earlier emails where you were describing your mother walking past you and not responding to your questions and comments it was like you were invisible) when I don't respond to you I imagine you feel similarly. I hope you can glean some understanding of how these dynamics play out in your significant relationships. Till we talk again. I'll be thinking about you.

Peter

The therapy continued for two months with Diana making considerable progress in her ability to use the therapist as a constant self-object. The decision to end the initial contract was mutual as Diana was beginning to feel more secure in herself and confident in her ability to move forward with things in her life that she had been putting on hold for some time. A year later Diana contacted the therapist saying she was very anxious about a new relationship and worried her old insecurities were beginning to get the better of her. After reading over their previous emails, it was easy for the therapist to pick up where they had left off. Over the following four weeks they were able to separate out Diana's issues from those of the newly developing relationship.

EXAMPLES OF CHAT ROOM THERAPY

Therapy in a chat room offers clients an excellent opportunity to better understand themselves. Clients can be shown over time how their ideas and beliefs either help them to function adequately in the world, or impair their functioning. The following brief chat example illustrates this issue. This client has great difficulty identifying and expressing feelings. She believes that one should keep one's feelings private and also that others cannot be used for support. These beliefs cause the client to be isolated from others and make it almost impossible for her to work through feelings.

Client: I don't know...I don't think I feel much like talking.... I feel kind of like...

Client: like I'm tied up...on the inside...

Therapist: *well, there's quite a value in sticking with that and talking anyway.*

Client: why?

Therapist: *because it's a valuable skill to learn how to talk when you don't want to,*

Client: oh

Therapist: *because that's precisely the time you need to.*

Client: oh

Therapist: *talk about the tied up feeling.*

Therapist: *ok?*

Client: about what?

Therapist: *about what you are feeling inside.*

Client: I kind of feel like I have no control over anything right now...

Therapist: *powerless?*

Client: yes

Therapist: *I don't like it when I have that feeling.*

Therapist: *talk more about this powerless feeling.*

Client: well

Client: sometimes I feel like others have more control over what happens to me than I do

Client: and not so much control sometimes...but

Client: circumstances...that happen

Therapist: *are some of those circumstances happening now?*

Client: I don't know...I guess so...to a certain degree

Over time, this client became more able to identify and express her feelings to the therapist and understand that it is entirely normal and natural to have uncomfortable feelings and talking about one's feelings to a trusted other can significantly help one manage the discomfort.

This next chat example illustrates the vulnerability and openness that this same client achieved with her therapist over time. She begins to talk about the despair and isolation she feels. Although the client and the therapist have discussed mandated reporting issues, she needed reassurance that she would be safe if she opened up about these very uncomfortable feelings and the therapist would not overreact to her discomfort. The client had been arrested for a minor misdemeanour and had not told her husband (or anyone else). She was facing a court date and all the issues and concerns that brings without any support, other than the online therapist. Her uncomfortable feelings intensified as the court date approached.

Client: ...can I say anything to you?

Therapist: *Just about. If you are going to hurt yourself or someone else, I would have to take some*

Therapist: *kind of action. Other than that, yes.*

Client: I certainly would never hurt someone else...

Therapist: *I have to report child abuse*

Therapist: *but that's about all.*

Therapist: *so say what you need to say.*

Client: maybe I better not

Therapist: *are you thinking about hurting yourself?*

Therapist: *?*

Client: no...I don't think so...but...sometimes...

Client: sometimes I wish I could go to sleep and either wake up and it's all over...or

Client: not wake up...

Client: but don't worry

Therapist: *that is not something I have to report.*

Therapist: *and I trust you – we've had several conversations about this.*

Client: yeah...I'm okay

Therapist: *it sounds like you are despairing.*

Client: Yeah...I feel like it

Therapist: *you're facing a lot of stress by yourself.*

Therapist: *and there are many unanswered questions.*

Client: yeah

Therapist: *but one of the worst things is you haven't been able to*

Therapist: *access the support you need from others.*

Therapist: *what keeps you from telling your husband?*

Client: I don't want him to know

Therapist: *will you be able to prevent that?*

Client: I don't know for sure...but I'm certainly going to try

Therapist: *you're pretty definite on that one it sounds like.*

Client: yeah

Therapist: *so are there others in your life? close friend?*

Client: I have a lot of friends...but none that I would tell that to

Therapist: *hmmm. I remember how difficult it was for you to tell me.*

Client: yeah it sure was

Therapist: *you thought I was going to run screaming from the chat room.*

Client: :o) lol

Therapist: *but I didn't.* :0)

Client: :o)

Therapist: *just want you to be ok.*

Client: me too.

The client was eventually able to develop a support system of women from the church to which she belonged. Her old beliefs concerning fierce independence and shame about feelings were replaced with an understanding of the appropriateness and need for support from others, and a greater acceptance of feelings.

CONCLUSION

There are many further issues to be considered in using online text within the counselling and psychotherapy profession. The Internet has forced businesses to rethink their previous bricks and mortar mentality. Mental health professionals and governing agencies historically have been cautious in their ability to assimilate change. Concerns for public safety and territorial battles over the economic and political landscape have had a hand in slowing this process. In addition, a great many issues remain to be addressed regarding international sanctions, legal and ethical issues, setting standards of practice and applicable certifications. Client safety and confidentiality is of paramount importance in the mental health profession. Making sure that we do all we can to protect the public is essential. No face-to-face or online system is going to ensure 100 per cent security and no client truly intent on harming themselves will be stopped. However, we must weigh the risks against the potential benefits when we consider this topic.

Developing appropriate models for working online is still in its infancy, and although it may be the case that a great deal of face-to-face therapy theory and practice is directly applicable to working online, more research needs to be done in this area to address the unique aspects of this practice. In addition, keeping up with improvements in technology will mean that the research will need to be a constant flow of updated material. Just as the Internet has given us tremendous opportunities for accessing knowledge and connecting with one another, the technologies of tomorrow will provide more possibilities and challenges. Continued efforts in this service are sorely needed to begin breaking down both the stigma of mental health as well as access to services.

REFERENCES

Cook, J. (1993) *The Book of Positive Quotations*, Fairview Press, Minneapolis, MN.

Seruya, B. (1997) *Empathic Brief Psychotherapy*, Jason Aronson, North Vale, NJ.

Stofle, G. (2002) 'Chat therapy'. In Hsiung, R. (ed.) *E-Therapy: Case Studies, Guiding Principles and the Clinical Potential of the Internet*. W.W. Norton, New York.

3 Conducting group therapy online

YVETTE COLÓN AND BETH FRIEDMAN

INTRODUCTION

With the availability of online forums for every mental health concern, more and more individuals are turning to the Internet for psychotherapy and support. There are many people who could benefit from a traditional psychotherapy group, but who would prefer to participate in an online group. In common with other applications of distance therapy methods, online groups have the potential to serve those who are unwilling or unable to participate in a face-to-face group. There are those for whom a face-to-face therapy experience is a geographical or physical impossibility, they may feel alienated from others or face possible stigma in seeking mental health services, or they may live in a rural community without readily available group therapy. In the online world, there are fewer limitations due to physical boundaries, and joining an online group can be a way to experience a clinical service in a direct, immediate and asynchronous way that adapts to the quickly changing nature of modern communication.

Some questions that need to be considered when looking at conducting therapeutic groups over the Internet are whether it is necessary to be in the physical presence of a group therapist to improve, whether a virtual community could be designed to create unconditional positive regard and whether strangers gathered together in an online forum are able to experience and convey empathy. Essentially, can a computer be the conduit for the emotionally corrective experience that is widely thought of as critical to psychotherapeutic change? An important feature of computer-mediated communication is the instant, 24-hour availability of the Internet. This allows participants to log in to their groups to connect with others and discuss whatever emotion, positive or negative, that they want to share and be supported in, serving to diminish social isolation, anxiety and depression (Finn, 1995). Another feature of computer-mediated communication is the idea of relative anonymity, facilitating self-disclosure. As a result, 'closeness' may be achieved at a safe distance and relationships among users can be strong, with many individuals finding the writing itself to be therapeutic (Murphy and Mitchell, 1998), offering an instant outlet for emotion, creativity and energy. In addition to factual information, group members can share photographs, poems, inspirational writings and even jokes to raise each other's spirits and provide comfort, with the ability to reread archived postings to get strength, support and information at the times when they need it most.

HOW ONLINE GROUPS TAKE PLACE

Online groups can be conducted in several different ways:

- *Usenet*: From the 1960s to the 1980s, prior to the creation of the Web, the first virtual communities online were text-based bulletin board services. One of the earliest forms of communication on the Internet was a discussion area called Usenet, a vast collection of 'newsgroups'. Peer support was available in newsgroups with the suffix 'alt.support' and thousands of people participated, and continue to participate in this mutual aid community. Usenet groups can still be found at Google Groups (http://groups.google.com/). A 20-year archive of Usenet messages is also available at Google Groups (http://www.google.com/googlegroups/archive_announce_20.html).
- *Bulletin board/message board*: A bulletin board is a program or location on the World Wide Web in which participants can read and write messages at any time that can be read by any other participant. The messages remain for the duration of the group, posted sequentially and usually organized by topic.
- *Chat group/chat room*: A chat group is a real-time exchange in which everyone is at their computer at the same time, 'chatting' with each other – much like a telephone conference call. At a predetermined time, the members of the group sign into the group using a special chat program so that they can communicate with each other. International chat rooms are also available through Internet Relay Chat (IRC); more information is available in *The IRC Prelude* at http://www.irchelp.org/irchelp/new2irc.html.

- *Mailing lists/listserv*: A mailing list is a private email subscription in which each subscriber receives a separate copy, via email, of each message that is posted, either in individual or digest form each day, week or month. Through these messages members can maintain ongoing communication with other list members who share a particular diagnosis or common concern.

THE GROWTH OF RESEARCH

Although we are approaching gender parity in the Internet population (in terms of numbers rather than level of participation), those who are underrepresented include ethnic minorities in most developed countries, large proportions of the populations of *some* less well developed countries, low-income individuals, residents of rural areas, individuals with less than a high school education and those over the age of 60 (Lenhart, 2000).

Stamm (1998) noted that mental health was ranked as the most common online telehealth consultation by the end of 1997, and mental health professionals have begun to explore the use of technology as a therapeutic and educational tool (see for example Miller-Cribbs and Chadiha, 1998; Giffords, 1998; Cafolla, 1999), but few studies have been rigorous evaluations. Meier (2000) states that little is known about the ways that online communication affect clinical interventions but, as Galinsky et al. (1997) acknowledged, technology was beginning to play a small, but important role in group practice. Their study was a survey of the nature of their subjects' knowledge, experience and comfort with technology-based groups, and they note that descriptive studies of computer-mediated support groups include interventions with carers of persons with AIDS, survivors of sexual abuse, Alzheimer's patients and carers and breast cancer patients. In a study about computer-mediated support groups, Weinberg et al. (1995) assert that although support groups can help people to increase their coping skills, only a small percentage of people in a crisis will choose to participate in one and that computer-mediated groups have the potential to serve those who cannot or will not participate in a face-to-face group.

Computer-based self-help/mutual aid groups for survivors of sexual abuse were described by Finn and Lavitt (1994), who found that these groups appeared to be of potential benefit to participants. They state that many of the therapeutic elements of group treatment, such as support, diminishment of shame and guilt, and acknowledgement of the universality of the experience, were present in the electronic medium. The anonymity of these groups was also helpful in minimizing the risk of personal involvement and dependency, although the authors continued to have reservations about the therapeutic benefits and went on to note that research about participants, benefits and potential harm of the groups is lacking.

Finn (1995) described a pilot project for an online support group and the advantages, disadvantages, potential for use and benefit to group leaders as well as participants, noting again that there has been little systematic evaluation of computer-based self-help groups and no research about their use as an adjunct to in-person support groups. Clearly, there is much quantitative and qualitative investigation to be done to assess the efficacy of online groups.

ASSESSMENT ISSUES

As with face-to-face groups, an online group member joins an open-ended or time-limited group to share difficult life experiences, seek understanding and support and seek help in making intrapsychic and personal changes. Their psychological status may vary – they may be functioning well or have an identifiable maladaption. The focus is less on common group goals, as in a support group, and more on individual goals for each member, each needing to be active and engaged in the group process. Many online group participants seem to form therapeutic relationships well, and develop their own creative tools such as specialized discussions and group process reviews to increase their level of trust and intimacy with others in the group.

In a professionally facilitated group, it is important to assess a potential member's readiness for group work, in order to increase the potential for an optimal experience for all group members. For example, one applicant, Joe, sent an email expressing an interest in joining an online cancer support group. The facilitator replied, sending information about the process for joining the group and a list of screening questions, and Joe answered quickly, writing that he was a young single man with cancer who lived in a rural part of the United States. The nearest support group was over 100 miles away and he did not have the means to attend the group consistently. Both his parents had died years before and his closest relative was his younger sister, whom he described as 'troubled'. He had been treated with surgery and chemotherapy, and stated in his email that the primary concerns that he wanted to explore in the group were coping with his isolation as a young cancer patient and maintaining a positive attitude in moving forward with his life. He agreed to the group guidelines, which included completing a pre- and post-group questionnaire.

His answers on the questionnaire indicated a significant depression, some suicidal ideation and serious pre-existing psychological problems. He sent another email to the facilitator outlining more clearly his depression and recent events, including self-mutilating and acting-out behaviour, and stating that he was also involved with a psychotherapist and case manager in the community support programme at his local mental health centre. The facilitator called Joe to discuss the email he sent and the issues he presented. He was friendly and verbal, but also tangential, with rapid and pressured speech. He was not able to focus on cancer-related issues; instead, he recounted his detailed history of trauma, isolation and chronic depression. He gave the facilitator permission to speak with both his case manager and psychotherapist as needed, and agreed with the facilitator's assessment that the online cancer patient group would not be appropriate for him at that time because his non-cancer related needs were too great to be addressed in such a group. Indeed, he did not have the ego strength or the ability to maintain firm boundaries to participate effectively in an online group.

FACILITATION

Online groups can be time-consuming for the facilitator and members because of the number of messages that must be read. The facilitator must decide beforehand how much time will be spent moderating the group, how available he or she will be, whether there will be any contact with group members outside the group sessions (via email or back channel communications, for example) and how emergencies will be handled. The seemingly limitless nature of the online world requires that the facilitator be aware of boundaries and limit setting with the group in order to avoid potential difficulties. The management of an online group requires not only clinical skills, but also some technical skills as well, as the facilitator must choose a chat, bulletin board or listserv program that he or she is comfortable with and can troubleshoot in case there are any technical problems. The facilitator must also decide if group members will be allowed to post or send messages without review or if all contributions will be reviewed and accepted/rejected, within the norms of that particular group and the parameters of standard group practice. An understanding of the advantages and disadvantages of computer-mediated communication is necessary, and the facilitator and group members must be able to type adequately, spell correctly and express themselves in writing well. Thought must be given to the security of the postings or messages in order to maintain the privacy and confidentiality of the group.

GROUP DYNAMICS AND UNDERSTANDING THE TEXT

Online support groups provide many benefits, such as the level of intimacy and trust being greater because participants may feel more comfortable disclosing and discussing their concerns. In synchronous online groups such as those in chat format, a skilled facilitator can overcome the tendency for group discussions to move quickly and sometimes randomly, by focusing the discussions and eliciting more in-depth information. In asynchronous groups such as those in bulletin board format, there are no time or space restrictions, as there are in face-to-face groups, so participants can send messages that are often more rich and meaningful, and can be examined more closely, than spontaneous verbal communication.

In online support groups, the facilitator and group members do not have the visual or verbal cues that are present in a face-to-face group. The facilitator must 'listen' in a different way by paying close attention to language, the way group members write, the ways in which they express themselves when they are feeling well versus when they are upset or sick. They might write in fragmented sentences or take less care with spelling and grammar when they are anxious, depressed or angry, or may not participate at all (and that 'silence' should have attention paid to it as well). The facilitator must be more active and sometimes more directive in the group to make up for the lack of eye contact and body language. If a facilitator does not send messages to the group on a regular basis, he or she may give the impression of being 'absent' from the group, making the group feel neglected by the facilitator, that he or she does not care or is taking the group less seriously, leading to a drop in participation.

The biggest challenge faced by an online group facilitator is the text-based nature of the group environment. The written word, in or out of the context of group communication, can be stark and direct, and humour and sarcasm can be misinterpreted easily and feelings can be hurt. In these situations, the facilitator must be a strong presence to mediate and guide the group through any conflict. If messages are not reviewed or rejected before they are posted or sent to the participants, the facilitator can use the conflict to further the group process by eliciting responses and exploration from group members (see case examples below).

Online group psychotherapy also raises the issue of members' commitment and responsibility. If the group is asynchronous, such as in bulletin board format, the perceived absence of others in the group and the knowledge that there is no one waiting to hear instantly about the members' concerns or improvement in mood or functioning can affect the motivation to participate and explore. In the online environment, text is the only mechanism to share and express emotion, and the members and practitioner alike must become adept at reading between the lines and searching for common themes and emotions. Members learn to communicate their feelings with each other, looking for cues other than the facial expressions and intonations of a face-to-face group. One important concept to explore when seeking clarity for therapeutic assessment online is pacing, as individuals have different expectations of how quickly someone should respond to a post or email. How should a facilitator handle group members who post less frequently to a support group? It could be a question of resistance, but it might also be individual style, lack of trust, or even simply a busy, tired or ill individual. It is helpful to set a group norm around this concept of pacing, and members should be given some suggestions, such as a weekly post requirement, to ensure commitment and continued participation. Members can have strong transference to the facilitator or other members when someone does or does not post or respond to another's comments. Even if a facilitator is actively reading and following the discussion, members may feel unheard or abandoned if no one replies to their posts. Pacing, in terms of the frequency of members' posts, how much is shared within each post, the free flow of ideas or the very sentence structure of a member's posts over time can also be used as an assessment tool.

For example, in one support group for family members of people with cancer, one participant's text communication style changed dramatically as her family member became more ill and her emotional state became more agitated. In this example, the client's writing style shifted from one of a well thought-out linear structure, to one of sentence fragments with more terse comments and ellipses between phrases, instead of commas or stops, as she became more angry and upset. Knowing how to distinguish this shift in style and using the written cues as a means to assess the client's altered emotional status was an important skill for the facilitator to possess.

Likewise, bulletin board support groups of an asynchronous nature, meaning that there is no set time when the group will meet, allow members to post when they feel most like sharing. With time to reflect on a post, members and the facilitator can think through a thoughtful, comprehensive reply, and the very act of writing can have therapeutic value and can be characterized as a therapeutic intervention, providing members with an immediate release of pent-up emotions as well as offering a vehicle for permanent change. The writing process may help a member to clarify feelings and find insight on his or her own as well as through the comments of fellow group members. Writing can lend itself to very powerful sharing (Murphy and Mitchell, 1998), and the facilitator can help members to recognize the difference between the manifest text, that which is directly stated, and the latent text which may be more subtle and touch on a particularly difficult feeling. By reflecting a member's words, the client may find increased insight and cognitive processing, through the sharing of a personal narrative, as found through journal-writing and narrative therapy.

While asynchronous groups offer a convenience and encourage participation when a member has free time or feels most like talking, members can also experience the group as distant, having a removed feeling because the facilitator and other members are not present with the client in a real, meaningful moment. However, mutual aid can occur dramatically and members can really feel a connection to one another, especially because members share when it is most salient to them. In this way, members often seem eager to open up and are surprisingly quick to discuss painful details that may take longer to share in face-to-face groups. In an online group for family members of cancer patients, one member found support and comfort when her fiancé died from a brain tumour during the course of the group. Cheryl, a 25-year-old woman, wrote:

> Dear friends, It is almost midnight my time. I got one of my friends to drive me over here so that I could leave a message for you all. I can't bear the thought of coming back and doing this while there are people in the building. Jack died at 2:15 (my time) today. He died at home. He was in a lot of pain. I am very sad. Have spent hours crying. Angry that I didn't get to see him again. Glad that he is free. I really don't know what to say. I think I am in shock. Please pray for me. There are so many feelings right now. I feel like a jumble of nerves. I keep feeling like there

> is something I should be doing, but I don't know what it is. I know you will understand.

Janet, another member, had seen her words and wrote back only hours later:

> Cheryl, I'm only home for a few minutes today - and something told me to check in - and I came right to your message as if by divine inspiration. I don't know if you'll see this, since you're heading out. First, I'm grateful that you found a way to share the news with us before you left. And that you knew we'd care. I'll hold you in my heart and in my thoughts. Be gentle with yourself this week. Love, Janet.

This example illustrates how empathy can transcend the written word.

CHARACTERISTICS OF GOOD GROUP WORK

There is a strong distinction between an unmoderated chat group or bulletin board and a private, professionally facilitated support group on the Internet. In order for good group work to occur, many of the same theories for face-to-face groups must apply. Similar to in-person groups, cohesion and bonding must occur for members to trust one another, and a skilled facilitator can ensure that the group is functioning well and moving towards mutual aid. Unless members become connected and care about what happens to one of their peers, they will not take that extra step of committing their time, energy and emotion to a group.

There are many reasons why online support groups can fail, and for the most part this happens when members are not invested in the feelings and outcomes of the other members. One of the most telling factors in successful online group functioning appears to be active group facilitation, and the role of the group leader seems most important at the beginning phase of the group. He or she must immediately establish clear group norms in order to relieve anxiety and encourage the group process of sharing to begin. At the outset of each group, members should be given a list of group guidelines, such as how often they are expected to post, how to share their concerns in the group and expectations for privacy and 'netiquette', a term that refers to online etiquette that includes issues such as how typing in all capital letters can be construed as shouting. Members should also be given instructions about how to participate, which need to be clear and simple, explaining the group process and technical issues for members with any level of computer or support group experience.

It is critical to begin to build trust during the beginning phase of the group. Questions arise about how to foster trust and cohesive group process in a therapeutic group without being in the same room. Just as in a face-to-face group, the facilitator must help members to trust each other and their facilitator, helping to define exactly how members can share their experiences in a respectful atmosphere. In support groups that have a particular focus, such as cancer support groups, members know why they are in the group, they share some common experience and they usually want information and relief from their anxiety and fears, looking to the facilitator for consistent messages. Good group work will entail opening up a conversation for more in-depth discussion and then looking for additional content from the members' posts and asking further questions to help elicit more shared discussion from the group.

Self-disclosure is essential for the group therapeutic process, and it is important for the leader to help to establish self-disclosure as one of the group norms but also allow members to share more personal issues at their own pace. The anonymity and absence of non-verbal communication seems to facilitate personal revelations and a more rapid development of transference feelings. Facilitators have often struggled online with what tone to use, and how much informative content to provide as opposed to staying with feelings through emotional content. Internet users are often active information seekers (Fox and Rainie, 2002), and they want access to information about their concerns. However, they can also have insightful discussions about their emotions online. A good facilitator can help members to do both, but the facilitator must make the group members comfortable and must anticipate possible transference feelings that may come up from very little information. For example, while in a face-to-face group the members see and hear the facilitator and each other, online they usually have only the first and possibly last names of the members. The lack of visual or auditory cues may make the therapist more mysterious or withholding, which could increase feelings of anger or frustration in the members. On the other hand, without seeing or hearing each other, members can concentrate on what a person is writing and possibly connect more deeply with others. These are important factors that affect how clients experience online groups.

CONFLICT

In online psychotherapy groups, the management of conflict involves a strong presence and leadership from the facilitator who can guide the group towards a resolution of the conflict and begin to address any longstanding issues that are stirred up. In one such conflict, a group member who felt misunderstood in her attempts to be supportive of another, addressed the bulletin board-based group in every discussion folder (the discussions were organized by topic in order to maintain some organization in the group):

> ROBERTA: 'All your words have been supportive of the stance I took with Jane. The thing that bothers me is WHERE WERE YOU WHEN IT WAS HAPPENING? I felt awfully alone in this situation. I *knew* I'd said /done nothing wrong. She lambasted me for no damned reason. But WHERE WERE YOU ALL? What I needed was the support of people here. Yet I said nothing. One of my biggest issues has always been an inability to say 'I need your help.'
>
> BECKY: Roberta. I'm sorry you felt alone in your argument with Jane. I was confused, afraid. I didn't understand what was happening. I really didn't want to intrude in the middle of it. I guess I was leaving it for someone like our group leader, who might have professional experience mediating a dispute, or to someone else here who could see clearly. I couldn't see the situation very clearly. My head was in the sand.
>
> ROBERTA: My responses to Jane were offered with care and sensitivity to those problems she exposed. Nothing was said in a disparaging manner. If anything, I see Jane's responses to ME as disparaging, ungrateful and totally without provocation.
>
> **FACILITATOR: I understand your viewpoint, Roberta, but I still don't see this discussion as anything different than previous postings. Your responses to Jane were made with care and sensitivity, as you say, but she's perceived them not at all that way. We can go round and round about who's right and who's wrong without ever getting to a resolution. Jane feels misunderstood, you feel provoked, where do we go from here?**
>
> ***SUSAN: I was sort of wondering the same thing as I've lurked the last several times I've logged on...where do we go from here?***

A few days later, Jane decided to drop out of the group, rather than work towards a resolution she was comfortable with.

FACILITATOR: I heard back from Jane and after much thought she has decided not to continue with the group. I expressed my sadness that she would no longer be with us, but I told her I respected her wish to leave. As of this moment, she no longer has access to the group. Well, this has been a most difficult experience, but I agree with Stan that we can do some good work here. I would like to continue our discussion about conflict in general, not necessarily about this situation in particular. Thank you all for your comments and feedback.

ROBERTA: I, too, am sorry Jane has decided to quit. I don't intend to discuss her decision, however, in her absence.

FACILITATOR: I sensed that others were hesitant to get involved with this conflict any more than they were. It's uncomfortable to watch/read/see something like that go on...After coming through our first difficult experience as a group, I'm wondering if you can all comment a bit more about your responses to what was happening at the end of last month. I noticed that a few people were hesitant to speak directly to the issue and others responded briefly. I was thinking about the dynamics of what was going on, but I don't think I can really do that unless I know a little bit more about how you all respond to conflict in general.

BECKY: Conflict makes me freeze up. I can't think rationally. I usually want to cry. I am completely unable to take sides, I just want the argument to stop.

SUSAN: I'm not exactly sure why, but conflict scares the hell out of me. I guess I'm afraid that it will turn violent or that I'll get deserted if someone is angry at me. It was always a threat growing up that if we caused problems, we would cause our mother to die. It was screwed up and painful.

STAN: Any time I have been in a conflict situation in the past I've had anxiety attacks of varying degree. I don't know why, but conflict scares me.

PHIL: Conflict doesn't mean having to resign. Why can't it be dealt with and resolved rather than running from it? Your pain has been evident, Roberta, what you might need from me (as I can't speak for anyone but myself) has not been clear.

In self-help/peer-led groups, where there is no professional facilitator, there can be conflicting roles and personal agendas. As a leader rather than a member, a professional facilitator can maintain a much greater degree of objectivity, make sure that all members are participating at a relatively similar pace, try to reflect common feelings and ask members to go deeper in a supportive way. As long as members feel 'safe' to share with people with whom they have common experience, they seem to foster their own positive environment. Especially in online groups where members can choose not to reply about emotionally laden content and shift the subject to lighter material, the facilitator can guide the discussion, pointing out differences, and continue to ask about more difficult topics if the group avoids emotional content for too long. One example of appropriate intervention might be around a topic such as spirituality, where a facilitator can recognize when a topic might be broached tentatively and open up a more generalized dialogue by asking more questions. In a support group for cancer patients, one facilitator wrote:

> **FACILITATOR: Several of you have brought up prayer in your posts as a way to cope. Does anyone want to share more about the role of spirituality in your lives? Have your beliefs changed since your diagnosis? How? Have you had any special experiences lately or connect with your illness in a way that seemed deeply meaningful or spiritual?**

TERMINATION

At the end of a support group, facilitators can choose to address the issue of termination through creative use of rituals, such as a group 'legacy' (a review of the group process and an opportunity to share what has been most helpful or meaningful in the group). Particularly online, the written form can offer a true legacy from their group experience, sharing final thoughts or a favourite quote or poem. This is a very special way for members to take something with them from the group that remains theirs to keep, unlike in a face-to-face group where members may not recall later what words they used to express their feelings. Online, members can print out their messages and keep them as a journal, and often these very specific rituals for sharing help clients to create an intimate and meaningful experience. In several online bereavement groups, for example, members chose to share through a termination ritual a particularly salient memory about their loved one who had died, and were also able to post pictures and share stories, helping the other members to carry the words of grief off the page in a deeply moving way. Through text-based communication, sharing can be deeply expressive and resonant.

The online termination process can continue to be a process for reflection, exploration and change. As they approached the end of another time-limited online psychotherapy group, the members began to say goodbye to each other and give feedback about their experiences.

CAROL: I think this experimental group of [ours] is wonderful and I'm truly honoured to be a part of it. Everyone here has touched me in one way or another. If there were some way I could crawl into the screen and hold each of you very tightly, stroke your hair, and tell you that you're really SUPER people, and that you're doing a great job dealing with your personal problems, I would. This is the only way I know how to do that. Here. With words...I wish you ALL a very happy, peaceful and emotionally comfortable new year. I love you all.

FACILITATOR: Lori also expressed her wish for more feedback from me. Here's the place to talk about it because I have some definite ideas about this, but I'd like to hear from all of you first.

LORI: As far as my wanting more interaction with [the facilitator] in here. That's my stuff. In groups, for awhile anyway, I like to have some guidance from a professional, I like to see some behaviour modelled, actually. Also, perhaps I felt that [the facilitator's] role here was a little ambiguous. There were times when I wondered what [the facilitator] was thinking.

FACILITATOR: Lori, you're right, I was a hands-off therapist. Many times I felt that you were all doing just fine among yourselves - exploring, challenging, supporting each other - and it would have been superfluous or intrusive (to me) to jump in and interrupt the flow. Perhaps it isn't the expectations or the process, but the word 'therapy' which conjures up so much. Maybe it's just that the word and the process don't fit well together. Other times I just hung back waiting to see what would happen, not wanting to take too much control. Sometimes it worked, sometimes it didn't. This is the way I work in real life. It's a good suggestion to have me explain this at the outset. I don't [explain it] in f2f groups, so it didn't occur to me to do it here.

CAROL: I entered this experimental online therapy group with very good intentions. I've already stated my disappointment as to the non-response to my posts. If, as you said, threads can change quickly, and the moment to respond has passed...then, I suppose, that's not what I'd like to take away with me from this group – passed moments, non-responses, no feedback.

FACILITATOR: Can you comment on the times people did respond to you?

CAROL: Is this become MY [discussion]? I think any answer to your question would be redundant...I believe I noted that Roger has been particularly verbal in response to all my posts. I'm not sure what you're requiring of me. What I'm saying is that I'm not sure how you wish me to respond to your question.

FACILITATOR: I'm confused, Carol. Are we keeping score about the number of times group members have responded to another's posts? Or are we paying attention to the quality of the support in this group?...One of my questions is: if you or anyone else presents themselves as a very strong person, a person who can be supportive and caring of others, a person who doesn't seem to need anything in return (for whatever reason), what happens when they realise they want something back and haven't let others know?

And finally, Carol's goodbye at the end of the group:

CAROL: When I typed 'join group' just now, I had the strangest feeling that I might receive a message like, 'Sorry, shop's closed!' Boy, was I glad I was wrong. The sensation wasn't a pleasant one...God speed. I'm a better person for this experience. Thank you.

During the group and even afterwards, members can choose to reread specific posts that provide comfort, either from another member or the facilitator, offering a 'holding environment' therapeutically for the client who is able to receive support whenever it is needed.

CONCLUSION

Online group participation and facilitation have many implications for research and practice. Examples of these are the exploration of the facilitator's interventions, participation and self-disclosures; how they can affect the group dynamics, and how the role of facilitator is perceived. We need to examine how trust is developed among group members, and another example of interesting research would be the assumptions made about factors like race/ethnicity by the group, and assessing socioeconomic background from writing style and use of colloquial phrases. Online group therapy can provide opportunities as well as challenges for the mental health practitioner. While the therapist should be cognizant of all the advantages and disadvantages, as well as any legal and ethical concerns, of providing group therapy online, this modality offers ground-breaking and satisfying opportunities for many participants to get help in making interpersonal changes and for professionals to infuse their work with creative, new skills and strategies in doing group therapy.

REFERENCES

Cafolla, R. (1999) 'An introduction to the Internet for independent group practitioners'. *Journal of Psychotherapy in Independent Practice,* **1**(1): 75–84.

Finn, J. (1995) 'Computer-based self-help groups: A new resource to supplement support groups'. *Social Work With Groups,* **18**(1): 109–17.

Finn, J. and Lavitt, M (1994) 'Computer-based self-help groups for sexual abuse survivors'. *Social Work With Groups,* **17**(1/2): 21–46.

Fox, S. and Rainie, L. (2002) 'Vital decisions: How Internet users decide what information to trust when they or their loved ones are sick'. *Pew Internet & American Life Project* <http://www.pewtrusts.com/pdf/vf_pew_internet_health_searches. pdf>. Issued 22 May 2002. Retrieved, 15 August 2002.

Galinsky, M.J., Schopler, J.H. and Abell, M.D. (1997) 'Connecting group members through telephone and computer groups'. *Health and Social Work,* **22**(3): 181–8.

Giffords, E.D. (1998) 'Social work on the Internet: An introduction.' *Social Work,* **43**(3): 243–51.

Lenhart, A. (2000) 'Who's not online: 57 per cent of those without Internet access say they do not plan to log on'. *Pew Internet & American Life Project* <http:// www.pewinternet.org/reports/toc.asp?Report=21>. Issued 21 September 2000. Retrieved 7/29/02.

Meier, A. (2000) 'Offering social support via the Internet: A case study of an online support group for social workers'. *Journal of Technology in Human Services,* **17**(2/3): 237–66.

Miller-Cribbs, J.E. and Chadiha, L.A. (1998) 'Integrating the Internet in a human diversity course'. *Computers in Human Services,* **15**(2/3): 97–109.

Murphy, L.J. and Mitchell, D. (1998) 'When writing helps to heal: E-mail as therapy'. *British Journal of Guidance & Counselling,* **26**(1): 21–36.

Shopler, J.H., Galinsky, M.J. and Abell, M. (1997) 'Creating community through telephone and computer groups: Theoretical and practical perspectives.' *Social Work With Groups,* **20**(4): 19–34.

Stamm, B.H. (1998) 'Clinical application of telehealth in mental health care.' http://www.apa.org/journals/pro/pro296536.html.

Weinberg, N., Schmale, J.D., Uken, J. and Wessel, K. (1995) 'Computer-mediated support groups'. *Social Work With Groups,* **17**(4): 43–54.

4 The supervisory relationship online

MICHAEL FENICHEL

INTRODUCTION

What exactly is meant by 'online supervision'? The answer to this question determines what the reader will be looking for in terms of information or assistance in considering how Internet-facilitated processes can and should be utilized. To begin with, the nature and focus of online supervision may encompass ongoing consultative support for clinical work that takes place primarily online, for face-to-face work in which the Internet is used primarily to access supervision, or in some combination. The Internet is a tool, perhaps even an environment, but not a process. Supervision is traditionally a process, over time, which in turn mirrors and directs another process, such as counselling or psychotherapy.

Without replicating the still-raging debates about 'online therapy' (Fenichel, 2000a) and whether online clinical work is simply a matter of transposing a traditional face-to-face activity into a text-based form, or an entirely new process, or even a new synergistic integration of processes and modalities (Fenichel et al., 2002), suffice it to say that supervision may be easier to define than 'therapy', but not necessarily. As with most clinical activities, context is critical.

In terms of asking why one would be particularly interested in engaging in supervisory activities online, there are several obvious and readily identifiable benefits. The first argument, one shared by proponents of online counselling, therapy, coaching, or 'e-therapy' (for example Grohol, 1998; Stofle, 2001), derives from some basic reality factors as discussed elsewhere in this book. Simply stated, Internet-facilitated communication allows practitioners, clients and supervisors alike ready access to one another, in some cases allowing competent and appropriate mental health services to be utilized by people who otherwise might not have anywhere to turn. The mental health practitioner who engages in online clinical work has as great a need for ongoing supervision as does the therapist engaging in traditional, office-based practice, if not more so. Moreover, there are undeniable and unique risks and benefits in providing such services online. The supervisory relationship, as in other counselling and psychotherapy endeavours, continues to be a hallmark of training as well as a valuable tool to ensure not only professional growth as a clinician, but 'best practice' skills which include a sensitivity to a range of ethical and practical issues.

SUPERVISION TYPES AND MODALITIES

Clearly, some elements of 'supervision' are readily recognizable across situations as well as modalities. Having an experienced person or group on which to rely in encountering a challenging situation, or even uncharted waters, is a hallmark of clinical supervision for mental health professionals. In some contexts, such as the workplace or education, supervision may entail legal or organizational obligations to ensure the appropriate actions of students, trainees or subordinates, while in many professional situations it is considered standard practice to engage supervision in order to not only develop and maintain skills, but to facilitate an ongoing process of skill-building and self-understanding. Indeed, in many countries, such as the UK, the main professional bodies require it for counsellors and psychotherapists as an essential element of good, ethical practice. No doubt there is a continuum of enlightenment across all these various activities, as supervision in healthcare, especially in its relatively highly developed form common in counselling and psychotherapy, is clearly a different matter from supervising a doctoral thesis, the kitchen in a restaurant, the delivery route for a newspaper publisher or classroom teachers in a school, to name but a few examples. Whether for accountability, best practice, or skill development for its own sake, supervision carries with it a social responsibility.

In psychotherapy outcome research (for example Luborsky et al., 1975), it is often concluded that one needs look at the match, or 'fit' (Fenichel, 1986) between the client and therapist, between the presenting problem and the type of treatment offered, and the 'non-specifics' such as comfort level, sense of warmth, empathy, genuineness and so on. Recent study has demonstrated that targeting goals and identifying expectations are crucial in being able to determine the course of treatment and assessing the outcome, and this is no less so in utilizing Internet-facilitated supervision strategies. Clearly, just as there are many types of counselling and psychotherapy, for many types of clients and therapists, with many possible presenting situations, there are many variables to contend with in terms of seeking or offering a supervisory relationship using synchronous or asynchronous text-based interaction rather than face-to-face discussion.

There are both new opportunities – such as the immediacy and new possibilities of online group consultation described in 'Myths and realities of online clinical work' (Fenichel et al., 2002) – and new challenges, such as finding a comfortable and supportive mode of online communication which is sufficient to meet the needs of ongoing clinical casework. There are also some 'trade-offs'. For example, one might bring case notes to face-to-face supervision sessions while one clearly needs to be cautious about sending such material online. On the other hand, if a client posts troubling messages, expressive artwork, or a self-revealing home page on the Web, it is an easy matter for a therapist to direct the online supervisor directly to those materials, rather than describing them or printing them out and bringing them to a supervisory session.

But one does not necessarily need to become involved in multimedia, multi-modality exploration; it is possible to utilize online supervision much like more traditional approaches: 'Dear supervisor, my client is beginning to express strong feeling towards me in ways which are not entirely appropriate, although expected.' In one scenario, this client might be displacing a reaction towards the loss of a loved one, or, in another, an infatuation with the therapist, or perhaps split-off anger at the world. A discussion might be very much like that which could happen in an office, or by telephone, exploring the client's life events and the process of therapy at that point in time. But another scenario might be one where the client is posting messages of love towards the therapist on a public message board, or sending dark and dismal email tomes to the therapist, or asking inappropriate questions, privately or publicly.

Recent study in the US suggests a growing crisis in the ability to provide graduate students with the type of clinical supervision that they seek in their advanced training. Rouff (2000) for example, describes how a confluence of factors, ranging from the restrictive 'managed care' environment prevalent in the US healthcare system, a decrease in the popularity of psychodynamic theories and reduced funding for psychotherapy training have made the availability of 'traditional' psychotherapy supervision more difficult to find. Still, even the most respected guides to clinical supervision, including recent exposition about the diversity of real-life clinical practice, do encourage supervisors and supervisees to maintain an awareness of factors that utilize traditional conceptions of human interpersonal relationships and self-understanding. Jacobs et al. (1995) noted that:

> Supervisees and supervisors these days are involved in discussing many kinds of treatment. They learn not only about insight-oriented psychotherapy but about crisis intervention, brief treatment, the management of disturbed behaviour, psychopharmacology and family and group treatments. Any of these treatments should be built on a psychodynamic foundation. Understanding a patient's conflicts and defenses, his values and ambitions, and his way of relating helps the clinician in developing a treatment plan that is both acceptable and meaningful to the patient. (p. 9)

It is difficult to ignore the fact that, like traditional therapies, supervision opportunities are increasingly more difficult to find in 'pure' form and the use of the Internet for self-guided and professionally facilitated supervision is dramatically expanding. Very few graduate programmes teach practical skills for text-based clinical work (or the potential of other online modalities). Continuing education for face-to-face practitioners on this topic is still in its infancy and professional organizations continue to debate hotly how or if online therapy can exist and be both efficacious and ethical. Like the earlier development of traditional supervision (first the self-supervision of Freud, then peers, then graduate and post-graduate training), there is a need to recognize that online services *are* being sought, some practitioners *are* effectively providing a wide range of treatment options that include or rely on the Internet, and the historical trend is repeating itself in demanding online supervision. Providers are appreciating the cost-effectiveness of treatment, consumers are turning to the Web for help with mental health just as they do for other products and services, and Web-based practitioners are developing skills and tool-sets based on experience, peer supervision and independent study.

Perhaps the first and most systematic effort to develop and describe peer-validated techniques and strategies for treatment across several types of presenting problem encountered online, using the various 'channels' and modalities that are accessible using a computer and a connection to the Internet, was pioneered by the clinical case study group (CSG) of the International Society for Mental Health Online (ISMHO). Co-facilitated by Suler and Fenichel (2000), a group of licensed mental health professionals, diverse in geography, discipline, training, and professional focus, has spent several years developing collective skills in evaluating initial 'calls for help', identifying effective and appropriate treatment plans, utilizing peer support for clients with special needs, and reinforcing our existing clinical skills while encouraging careful use of new opportunities, such as combining face-to-face sessions with an online support group, encouraging social skills through online activities which can then generalize to offline relationships, using diaries, art, Web pages and so on to encourage self-analysis and self-expression, and utilizing synchronous communication online to successfully avert crises and/or make appropriate referrals, sometimes to healthcare systems in another country. None of this would be possible without the immediacy afforded by the Internet and a network of peer support and supervision which remains in place 24 hours a day.

The CSG recently explored ten myths of online clinical work (Fenichel et al., 2002) and among the myths was the idea that text-based communication is somehow an inferior or inadequate means of expressing human experience. However, Shakespeare and Freud, to name but two well-known writers, knew differently. The word is powerful. We found that online relationships can be, too. Clearly the supervisory relationship is no exception.

TRANSFERENCE AND COUNTERTRANSFERENCE

As was seen among the members of ISMHO's CSG, it is also a myth that in online therapeutic relationships one cannot or does not develop strong feelings much like traditional transference relationships and countertransference as well. Because of this fact, it is important for the online supervisor to understand and acknowledge the validity and intensity of such therapeutic relationships. After all, the supervisor may take on an equally important, and real, role for the practitioner, especially if the former is one who is likely to be available for quick response should there be need for consultation due to an emergency or a crisis. It is also important for the online practitioner and supervisor to recognize the many possible meanings and variations of online communication styles, ranging from impulsive response styles, to email 'silences', to use of colour and font, to violation of the therapeutic frame or pre-established boundaries, such as pursuing access to the counsellor outside the times and modalities established under the therapeutic contract. Such considerations are particularly important when working in a theoretical model where the relationship is a part of the data to be highlighted within the ongoing work, but are liable to be important for those working in a number of other supervisory approaches too. In looking at the supervisory relationship in relation to a client's pattern of interpersonal relationships, an exploration of dynamics such as respect for personal boundaries and preferences can play an important role in assessing, identifying and working with key therapeutic issues.

case vignette

One member of the ISMHO CSG had a face-to-face client join the bulletin board that she moderated on her website. This resource not only helped the client to make the transition from discussing her feelings and problems with only her therapist to discussing them with others through the message board and later with people in her life, but observing the client in this medium gave the therapist new and valuable information about how her client interacted with people. Issues of transference and countertransference arose in this context that prompted the therapist to seek online supervisory input. For example, when the client became jealous of the attention her therapist gave to other users of the same website, she voiced this by posting angry messages about the therapist in the public forum. Because the client's words were posted online, the therapist was able to copy and paste the post verbatim and send it to the CSG's e-list to receive collegial feedback and support almost immediately. This could be done not only before responding appropriately to those at the website message board, but also before exploring these feelings in the face-to-face therapy (Fenichel et al., 2002).

There are also therapeutic models that rely less on the therapeutic relationship and more on self-generated changes in cognition and behaviour, guided by a therapist using any number of strategies ranging from rational-emotive behaviour therapy (see Fenichel, 2000b) to use of diaries, narrative analysis, role-plays, bibliotherapy, community-based behavioural 'homework' assignments, self-help and so on. Here, as mental health professionals are just beginning to explore the ways of integrating such approaches into online/offline treatments, the clinician is often clearly treading on new ground, and thus the supervisor needs to be aware of both the theoretical frameworks being utilized and also the nature of online clinical work generally, and the specific strategies being employed by the therapist. For the supervisor, it is important also to be aware of a number of factors which impact the process and efficacy of the therapeutic endeavour, ranging from the clinician's and client's ability to 'transparently' utilize the technology without it getting in the way, to familiarity with legal and ethical issues (such as privacy/confidentiality), to having the flexibility and pragmatic outlook which allows for systematic review and refinement of the process, treatment parameters and movement towards attaining specific goals.

It should also be noted that some of the 'classic' supervisory approaches towards psychotherapy are based on the review of session notes or transcripts. In using email (and in some cases chat transcripts) there is a natural opportunity to review both process and content. When used in a peer supervision process, such as utilized by ISMHO's CSG, a consensus often developed about key issues and strategies. This, of course, is consistent with what has long been known about effective case supervision in traditional (face-to-face) contexts, such as in the findings of Goldfried et al. (1998) which found that regardless of theoretical orientation, the opinions of 'master therapists' converged on what they felt were the most significant events when shown session tapes.

NEW MODELS, OPPORTUNITIES AND CHALLENGES

A recent three-year report by the ISMHO CSG (Suler and Fenichel, 2000) vividly discussed the case of a military pilot who sought online support for stress and anxiety relating to identity issues, work relationships and sexuality, but who was also geographically accessible. The client found an online therapist who happened to be an expert in gender identity and also very experienced in a variety of online activities, ranging from chat to email to utilizing a message board around which a thriving support community evolved. This was clearly a perfect 'match' for this client who was constantly mobile but often in need of support and motivated to engage in a therapeutic process. The benefits were clear to both therapist and client. This therapist, in a professional and ethically sound manner, sought peer supervision for this case through the ISMHO CSG, consisting of other experienced, licensed, *online* mental health professionals. She credited this peer supervision support – which could be found seven days a week, virtually around the clock given that the group members were located across several time zones – with providing valuable guidance and validation for her efforts.

Moreover, this peer group included other members who were experienced not only in psychotherapy practice (online and offline) but some who also shared the therapist's expertise in gender/sexuality issues and use of the various modalities of Internet-based therapeutic work. In a great many ways, having a diversity of experience was a cornerstone of responding with appropriate suggestions and concerns regarding presenting problems as they were first encountered by the group's clinicians, and also as a means of support and affirmation during longer term treatments which entailed twists and turns of events. Just as the diversity and vastness of cyberspace may bring to therapists a wealth of new acquaintances, social and cultural situations and novel experiences, the potential for peer supervision as well as other types of graduate training and continuing education is clearly enormous.

When one can draw upon colleagues with expertise in areas ranging from addictions to crisis intervention, gender issues to existential interpersonal angst, and with familiarity across multiple cultures, treatment modalities and national healthcare systems, the opportunity, responsibility and challenge of having such a network of peer support and supervision cannot be understated. Still emerging through experience, consultation, research and collaboration, the models being seen among individual practitioners as well as online 'virtual clinics' continue to become more versatile, more grounded in the types of clinical presentation which online practitioners actually experience and more creative in utilizing the opportunity which presents itself through the use of technology and Internet-facilitated communication. However, while young children and teens are already computer-savvy and inclined to seek information and help online as a first resource rather than as something obscure and difficult to do, it is clear that the teaching of online skills – not only therapy/counselling but also basic communication, perception, behaviour and phenomenology – is still only slowly emerging at the undergraduate and graduate level. As in the initial days of psychotherapy, only a few pioneers are striving to lay out models to understand and facilitate systematically the positive aspects and dynamics of online clinical work, or to tap into the power of human connectedness in order to support the types of 'calls for help' which are commonly seen by online clinicians. There is no doubt that the supply of knowledgeable generalists and specialists who are able to provide competent supervision of online clinical work will need to expand greatly in order to meet the demand.

One effective model, utilized by the ISMHO CSG and described in detail by Suler (2001), uses a closed-group email list. The process combines both traditional group structure, beginning with a framework emphasizing confidentiality plus ground rules for presenting cases and participating in a productive manner for the benefit of both clinicians and clients, and also elements of traditional clinical supervision such as emphasis on presenting problem, determining suitability for treatment (Suler et al., 2001) and pursuing a treatment plan which appropriately reflects shared goals and expectations. Suler (2001) also notes that:

> As in all groups, the skills and personalities of its members greatly shape the group experience. To help maximize confidentiality, cohesion, and group identity in the ISMHO case study group, we limited the number of members to approximately 15 and kept the group closed during each round of case presentations...Equally important, the founders of the group must be clear about it's purpose and philosophy. Exactly what kind of clinical work will be discussed? Will it focus on any particular theories or approaches? What values about psychotherapy and clinical supervision does it uphold? Of course, these questions pertain to any clinical group, online or off.

When the overriding shared value became 'what is the most appropriate intervention to assist the client in moving forward in a positive or more self-knowledgeable direction?', the group as a whole tended to serve as a dynamic and supportive forum for helping to direct the practitioner in decision-making during critical junctures in treatment, often while in the midst of an 'asynchronously live' case presentation (presented very close to 'live' with reports presented as the process was still in progress, but because they were sent on an email list, and readers could 'time shift' by accessing the messages with a delay, it was also considered 'asynchronous' rather than a completely live, real-time or 'synchronous' experience). Very quickly, because of the relatively small group size and likelihood of subgroups developing as expertise related to each case became apparent, while there were designated co-facilitators who had been present since the beginning of the first group and through each successive year's cycles, the group became a cohesive and trusted source of peer supervision as each member developed trust and learned each other's areas of strength and experience, while the group as a whole was 'there' for each presenter as cases were discussed and well beyond.

Practical considerations for utilizing this model of case supervision with peers extend beyond the structure to both human and software considerations. First of all, in order to have a comfortable, 'transparent' environment where one could freely communicate with a minimum of frustration due to 'technical difficulties', an easy-to-use email list was set up where, once subscribed, members could reply and post quite readily, posts were distributed almost instantly (which was often crucial when an emergent problem was being presented) and files such as case transcripts or relevant articles could be shared privately, in accordance with the ground rules established at the beginning. Of course, as in all case study endeavours, client confidentiality and informed consent were paramount.

Apart from structural and case-relevant considerations, Suler (2001) also describes a number of aspects of which those new to online small-group discussions need to be aware. His ten-point organizing 'rules' for effective email list peer supervision discussions begins with such things as having an awareness of how 'replying to' messages on the listserv entails the message being sent to everybody on the list and extends into more experiential phenomena such as:

- Understanding how list members may vary in their *pacing* (both receiving/processing and replying to mail)
- Being precise in quoting previous text, and in stating a comment or question
- Learning to accept and anticipate the ambiguity which often accompanies online communication, due to absence of facial and other non-verbal cues
- Being professional, respectful and supportive in tone, not critical
- Letting the group know of concerns, pending absences and so on
- Having the means (for example through use of a 'process' thread) to step back and clarify a process issue in the group, or confusion as to a communication.

The CSG, like other supervision groups, benefited from having a structure that established basic requirements, rules and timelines. But unlike many other groups, there were a variety of ways in which therapists were working with clients (for example combinations of synchronous and asynchronous communication, chat and email and face-to-face supplemented by message boards and email, to name a few). By the same token, this allowed, over a two-week presentation period, unconstrained by the clock time, for participants to present case material derived from a number of modalities. Some presentations read like traditional case transcripts and were taken from email, while others provided material from message boards and chat sessions, depending on the therapeutic relationship. Some presentations were concise, others offered a great deal of material. Some presentations were focused on specific questions, such as whether the therapeutic focus was being 'correctly' placed or not, or whether a face-to-face referral was more appropriate than online treatment, or how to react to what appeared to be an imminent risk and 'call for urgent help'. Still other presentations encompassed more ongoing, theoretical concerns, such as how to integrate self-help and guided online exploration with more traditional client–expert relationships, or how to better integrate one's website offerings with one's clinical work. Of note, colleagues continued to be available for consultation after formal case presentations were completed, often offering a taste of what might be expected in cases of much longer term treatment and supervision.

Increasingly, peer supervision has been a valuable and often crucial support for the growing number of online therapists working with a wide variety of clientele, predominantly utilizing email, but also drawing on technology's power to facilitate human interaction using chat, message boards, video, instant messaging and a wide variety of expressive outlets which can be brought to bear in combination of the client, therapist and the connective technology of choice.

Apart from the case of a military person on the move, and clients who fear the stigma of presenting face to face for specialized treatment, the types of case which the ISMHO CSG worked with during 2001–02 also included gender/sexuality issues (where the group had several resident experts to provide analysis and offer suggestions and resources); boundary/relationship issues (where the therapist worked with the client across message board, email and telephone modalities and relied on peer supervision at times when the client made dramatic statements or gestures); and a call for help from a college student suffering panic attacks and major depression who sought treatment under a nationalized healthcare plan (which his therapist learned about and arranged through consultation with peers in that country through the CSG).

Similar to the office, hospital or university-based supervisor who is engaged to support the practitioner who encounters a diverse assortment of clinical cases, it is an extraordinary responsibility and challenge for any one person to knowledgeably and effectively provide supervision across the full range of situations a supervisee might present. If this is so in face-to-face supervision, where the supervisor often has more direct access to case records and certainly more direct (face-to-face) contact with the supervisee, imagine how resourceful and experienced the supervisor of online clinicians needs to be. With this in mind, it is felt that the power of peer supervision such as the ISMHO CSG has been providing is quite profound and more effective at being able to respond to the types of presenting problem and circumstance which the online clinician is likely to encounter. However, that said, there are obviously a great many situations in which one might well seek individual supervision as well, particularly for specific cases or circumstances, small online practices, staff development or training purposes and skill-building of leaders within specialized online communities and mental health websites.

As mentioned in the introduction to this chapter, online supervision can mean a number of things. One possibility is clearly that a supervisee seeks the experience and guidance of somebody skilled in a particular type of therapy or disorder or client population, and face-to-face access is not a possibility. In this case the client and therapist may work exclusively face-to-face, while some or all of the supervision may be facilitated by using Internet-based communication. Historically, telephone supervision has been used for similar reasons, without the mystique or controversy now seen surrounding the Internet and online therapy. Historically, telephone hotlines have been a source of immediate help for crises and those receiving the calls rely on round-the-clock supervision just as is available online. Supervision in clinics and graduate schools may be entirely oriented towards traditional therapy with students and adults who are no longer living traditional lives, in that they engage in communication, relationships and help-seeking via the Internet, and may talk about this in face-to-face therapy sessions, even as online counselling may entail discussion about face-to-face relationships. Online life is becoming integrated into offline life and vice versa. Supervisors and therapists need to know this and make accommodations to understand the power of the Internet, as well as to grow beyond the walls of the office setting. An individual therapist or supervisor may be quite knowledgeable about some aspects of online life, cross-cultural issues, practice regulations, psychotherapy techniques and the ethics of working within his or her own scope of knowledge and competency. But when the entire world has access to one's online office, and when clients seek help and online practitioners seek guidance and supervision, the power of a group such as ISMHO's is immeasurable, drawing on international, cross-disciplinary and multicultural perspectives shared within the group as well as the experience of the group's leaders and individual members as it relates to the individual 'case' at hand. Increasingly, we live in an interconnected world, and an ever-expanding community of mental health professionals may have similar experiences worth sharing.

One altogether different type of online supervision also needs to be mentioned, and that is supervision of mental health practitioners, facilitators and hosts of online communities. Many very large communities, through official sponsorships of the provider, organizational support and/or legacy and word of mouth from community elders, take great care in providing training and mentorship opportunities for new community guides and leaders. While commercial interest rather than mental health may be the primary focus of sponsors' efforts at community-building, and while the hosts may not provide 'clinical supervision' per se, it is likely that mental health communities in particular will benefit in the near future from utilizing (paid) consultants who can facilitate group processes and the development of online social skills, as well as the professional development of community leaders, so that there is a better ability to identify and respond to calls for help and other serious situations which may arise. Similarly, as educational organizations and corporate healthcare organizations expand Internet-based benefits and activities, online supervision and continuing education may be extended into new forms of mentor relationships with faculty and educational leaders, corporate management and employee assistance programmes, as is already happening around the globe.

CONCLUSION

Notwithstanding one's competency to address myriad potential types of casework and work within one's area of knowledge and expertise, one needs to be aware of several practical concerns. First is the issue of the purpose of supervision. For purposes of licensing and accreditation or completing an advanced degree, online clinical supervision is not yet widely recognized as being acceptable. As with 'therapy', arguments about jurisdiction and definitions are rife and perhaps years away from resolution. There is little in the way of evidence-based support for the treatments now offered, beyond anecdotal reports.

Yet increasingly consumers are inclined to seek not only information, but also help, online. Even the conservative *Wall Street Journal* proclaimed in a headline (4 June 2002), 'Online Therapy Goes Mainstream', while noting concerns about privacy, confidentiality and the possibility of finding 'unqualified or unscrupulous counselors'. Thus, one admonition to the would-be supervisor of online practitioners, in addition to knowing the potential and social norms of cyberspace, is to 'know your supervisee', and by the same token supervisees might need to ensure that they 'know their supervisor'. Of course it goes without saying that for online therapists, ethical issues such as confidentiality co-mingle with some additional concerns, such as being sure one knows the identity of who one is working with – of great concern given the potential for minors writing in, a spouse posing as husband or wife of the other, or a client being at risk of harming themselves or others.

While there is considerable debate over special circumstances where anonymity may be essential for the client, there are also legal and ethical guidelines that bind the mental health professional. Beyond the ethical issues are some very practical issues, such as understanding the results of the 'online disinhibition effect' (Suler, 2002), and the various limitations that text-only places on our ability to accurately convey and understand wit, nuance, sarcasm and tone. For these reasons and others, not only does the online practitioner need to be knowledgeable, comfortable and competent with online treatment parameters, but also the supervisor must be even better versed in such important matters. This, again, is an excellent argument for having a peer supervision model such as described above, where one has a range of input from a variety of mental health disciplines and geographic/cultural/political contexts.

If one is to supervise even the most traditional therapy cases in the twenty-first century, one will inevitably need to be at least familiar with common life experiences such as found in chat rooms and emails, to be able to separate out distortions and fantasies from reality, as well as to empathize with clients' everyday lives. If one is then to work with therapists who themselves rely on Internet-based communication for clear, therapeutic work with clients, one is doubly obligated to know one's client, while also knowing the myths and realities of online clinical work.

REFERENCES

Fenichel, M. (1986) 'Patient–therapist fit': The relationship between congruence of initial treatment expectations and the outcome of psychotherapy. Adelphi University, Institute of Advanced Psychological Studies, *Dissertation Abstracts International*, **47**, 4.

Fenichel, M. (2000a) 'Online psychotherapy: Technical difficulties, formulations and processes'. [Online] Available: http://www.fenichel.com/technical.shtml.

Fenichel, M. (2000b) APA Convention Report: Aaron Beck and Albert Ellis. [Online] Available: http://www.fenichel.com/Beck-Ellis.shtml.

Fenichel, M., Suler, J., Barak, A., Zelvin, E., Jones, G., Munro, K., Meunier, V., Walker-Schmucker, W. (2002) 'Myths and realities of online clinical work', *CyberPsychology and Behavior,* **5**(5) [Online version: http://www.fenichel.com/myths/].

Goldfried, M.R., Raue, P.J. and Castonguay, L.R. (1998) 'The therapeutic focus in significant sessions of master therapists', *Journal of Consulting and Clinical Psychology,* (66) 803–10.

Grohol, J.M. (1998) 'Future clinical directions: Professional development, pathology, and psychotherapy on-line'. In J. Gackenbach (ed.) *Psychology and the Internet, Intrapersonal, Interpersonal, and Transpersonal Implications.* Academic Press, San Diego, pp. 111–40.

Jacobs, D., David, P. and Meyer, D.J. (1995) *The Supervisory Encounter.* Yale University Press, New Haven.

Luborsky, L., Singer, B. and Luborsky, L. (1975) 'Comparative studies of psychotherapies: Is it true that everyone has won and all must have prizes?' *Archives of General Psychiatry,* (32): 995–1008.

Rouff, L. (2000) 'Clouds and silver linings: Training experiences of psychodynamically oriented mental health trainees', *American Journal of Psychotherapy,* **54**: 549–59.

Stofle, G.S. (2001) *Choosing an Online Therapist: A Step-by-Step Guide to Finding Professional Help on the Web.* White Hat Communications, Harrisburg, PA.

Suler, J. (2001) 'The online clinical case study group: An email model', *CyberPsychology and Behavior,* (4): 711–22 [Online version: http://www.rider.edu/users/suler/psycyber/casestudygrp.html].

Suler, J. (2002) 'The online disinhibition effect' [Online] Available: http://www.rider.edu/users/suler/psycyber/disinhibit.html.

Suler, J. and Fenichel, M. (2000) 'The online clinical case study group of the International Society for Mental Health Online: A report from the Millennium Group' [Online] Available: http://www.ismho.org/casestudy/ccsgmg.html.

Suler, J., Barak, A., Chechele, P., Fenichel, M., Hsiung, R., Maguire, J., Meunier, V., Stofle, G., Tucker-Ladd, C., Vardell, M. and Walker-Schmucker, W. (2001) 'Assessing a person's suitability for online therapy', *CyberPsychology and Behavior,* (4): 675–9. (See Correction, *CyberPsychology and Behavior, 5,* 2002, p. 93).

Part II
Telephone and video links

5 Telephone counselling and psychotherapy in practice

MAXINE ROSENFIELD

HISTORY

It seems incredible that as recently as 1997, some practitioners regarded therapy by telephone, anecdotally, as 'not real therapy', and even with some derision. Yet today, many practitioners use the telephone for some or all counselling and psychotherapy sessions, and also for supervision (see Chapter 7).

The use of the telephone for help in times of crisis has been acknowledged for over 40 years. The issue for many in the counselling and psychotherapy world was the precise nature of the support that could be offered by telephone. Rosenfield (1997) defined the spectrum of help that might be offered by telephone and showed the position of counselling within that spectrum. The issues raised by some practitioners were concerns about how a client could receive therapeutic help, develop an ongoing relationship with the counsellor, ensure their location for each session was private and pay for the session if there was no face-to-face contact. Given that similar discussions now take place regarding the use of email and other technological innovations, it would seem that, as a profession, we tend to resist the new until external pressures, such as public demand, dictate otherwise.

WHAT IS THERAPY ON THE TELEPHONE?

Telephone therapy is best described as an ongoing, contracted relationship between the practitioner and client (or clients if a telephone therapy group is established) carried out entirely by telephone. Although some practitioners use the telephone for some sessions, this chapter is written in the context of all sessions taking place on the telephone with no face-to-face contact between practitioner and client. The contracted nature of the relationship sets telephone therapy apart from the work of telephone helplines that typically use counselling *skills* in one-off sessions, to support callers as a 'listening ear', perhaps offering information or referrals to other agencies, usually without exploring deeper underlying issues that would require further sessions and more in-depth work.

Most people in developed countries have access to a telephone and are familiar with using it. There is no doubt that the telephone is perceived to be a safe, confidential environment and that people appreciate the privacy it affords, making it a popular medium. The Telephone, Information, Support and Counselling Association in Australia (TISCA, 2000) calculated that in 1999–2000 a call was made to such a service every three minutes, in the state of New South Wales alone. A 24-hour counselling service for teachers in the UK received an average of one call every two seconds, every minute of every day over a 12-month period (TBF, 2001).

There are many reasons why a client may find that therapy by telephone suits them, in common with other means of distance provision, such as:

- If a client is feeling vulnerable (perhaps has been the victim of rape), does not wish to go out and meet with a 'stranger', or visit an unfamiliar environment.
- A client with mobility problems might not wish to travel regularly to a face-to-face session or might not find a practitioner operating within an accessible location.
- A client might have a specific need, be seeking to work with someone from a similar cultural background or wish to work with a specific practitioner who might not live nearby or even in the same country.
- Attending regular therapy when the nearest town is some distance away (assuming there is a practitioner there) makes the telephone a far more suitable medium for people in rural areas.
- Clients may not wish to go to a local practitioner, preferring instead to work with someone who they are unlikely ever to meet.
- For many people, having to travel to a practitioner's office, spend an hour or so there and travel back makes face-to-face work prohibitive in terms of time and possibly money.
- For clients who have a chronic, debilitating or degenerative illness, the telephone can enable them to work therapeutically where they might sometimes be too unwell to go to a practitioner's rooms. In

addition, it may be easier to use a telephone than it would be to type on a computer keyboard.

- The immediacy afforded by the telephone enables contact to be made without time delays and at the convenience of both parties.
- The telephone also provides clients with a degree of safety and control they do not normally have. This is discussed further below.
- The telephone provides an intimacy as the counsellor and client speak directly into each other's ears. It is possible to feel cocooned by holding the telephone to one's ear, which enhances a sense of safety and trust.

Practitioners must work somewhat harder than in face-to-face work to convey in their voice, words and contract the values and principles of their work. The obvious lack of eye contact and visual clues and cues means that therapy by telephone is heavily reliant upon the listening skills of the practitioner in particular, in order to be able to note and use in the sessions the variations in the client's voice, speech patterns, choice of words and so on. It takes some practice for practitioners to learn to trust their 'inner ear' totally on the basis of what can be heard and indeed some practitioners find they are not comfortable with the medium and working with so many unknowns.

Some practitioners, for whom seeing the client and observing their body language is an essential component to their work, find they cannot concentrate fully on the client when sitting alone in a room with a telephone for up to an hour at a time. There are those who do not like holding a handset for so long or wearing headphones. However, when engrossed in a session, most people do not notice that they are holding a handset for such a long time. Indeed, if they find themselves thinking about the handset or their seating position, it could be useful to reflect on what is happening at that point in the session. It is not ideal to work with the telephone on 'loudspeaker' as an alternative, as the sound distorts and this affects the sound of the client's voice tones.

Certain therapeutic orientations are well suited to the telephone. Cognitive-behaviourist approaches and person-centred (experiential) counselling are easily adapted to telephone work, although many eclectic practitioners could enjoy working in this medium. Gestalt and other techniques requiring practical activities are clearly not suited to telephone work (although some therapists may use the telephone for an occasional session).

ASSESSMENT SESSION

During the assessment session both parties can decide if they wish to work together and then agree to a contract. This will include the frequency, length and time of each session and the number of sessions before a review. There will also need to be some discussion about both parties' responsibilities in the relationship and the confidentiality bond, which includes instances when the practitioner might break confidentiality. In addition, the method of payment and when payment will be made need to be agreed. Fees are generally the same as they would be for face-to-face therapy, the cost of the telephone call made by the client equating in principle to the cost of travelling to a counsellor. Some counselling services operate with a freephone number, in which case the client does not have to pay for the call, but often the client pays, even if the practitioner is working overseas.

Although clients may request it, I do not send a photograph, explaining that a key element to working by telephone is not to have the visual image of each other, but rather to work only with what we say to each other, although at times it may become appropriate to explore images and fantasies. It can be likened to the position of a radio announcer whose image is constructed by the listener's mind. How often has the comment been made on seeing a photograph of a radio announcer, 'I never thought she'd look like that'? However, this is an area where practice varies. Some practitioners report that their clients tell them they place the counsellor's photograph near the telephone when they are having a session and it helps them to talk. At times I will ask a client to describe to me where they are sitting and how they are sitting and if they ask I will tell them about my office. I have found that this seems to settle a nervous client. Sometimes a little imagination and visualization can go a long way to helping to create the working relationship.

A date will be set for the first counselling session as the assessment and contracting draw to a close. In the event of working across time zones, the practitioner must be very clear about the time of the sessions from both parties' perspectives. After the assessment session, there may be a 'cooling off period' for both parties, set as up to five days after the session, where either can decide not to proceed, giving enough notice for the practitioner to use the time that was allocated for the sessions in other ways. Many practitioners offer the assessment session free of charge, with the exception of the cost of the telephone call. Some practitioners decide to make the call so that the potential client has no expense to bear, others prefer the client to call.

Often the practitioner will send a copy of the contract by post or email to the client so that both parties can be clear about what has been agreed before the formal therapy sessions commence.

During the assessment session, the practitioner has to be able to convey to the client that they will be receiving a professional service and should be prepared to answer the client's questions. These may be focused on the methods of counselling or on the client's perceptions of the counsellor – such as age, racial background and expertise in the area for which the client is seeking help. A practitioner working by telephone has to be prepared with answers to direct questions – the medium lends itself to 'bravery' on the client's part ('I can't see her/him so I won't feel judged and I can always hang up if I am not satisfied'). If both parties decide to continue, the sessions will begin, ideally one week apart, at a regular time and on the same day each week.

FREQUENCY AND DURATION OF SESSIONS

These will depend on the practitioner's style and the nature of his or her work, but in general a minimum of six sessions on a weekly basis is a useful initial contract to make, as that enables a good relationship to be built. It may be apparent that six sessions is enough or a further contract can be arranged. Session length, which can vary according to the practitioner's style, should be agreed at the assessment session. Each session should be of the same length so that both parties can work within that clear boundary.

TRUST

With no face-to-face contact at all and only a conversation or two and the voice he or she has heard, a client places enormous faith in his or her chosen practitioner. A practitioner must acknowledge this and ensure that his or her working environment is quiet, that he or she answers the telephone directly at the time the client is due to call and does not let an answering service or anyone else pick up the call first. The use of mobile phones is not to be recommended for their lack of reliability and privacy.

A trusting relationship can develop very quickly with telephone counselling clients and, anecdotally, it seems that fewer sessions are needed than in comparable face-to-face work. It would appear that the relative anonymity afforded by the medium enables clients to take risks and talk more freely sooner than they might in other settings. This also enables the practitioner to take risks and use all intuitive streaks before he or she might have done if working in the same room as the client. I have found that the disclosure of very painful events or experiences often happens within the first two or three sessions by telephone, whereas I found it took several more before a similar level of disclosure was reached face to face. Counsellors working by email report similar experiences, which again suggests the privacy that the medium provides is a positive factor in enabling clients to feel free and safe to disclose (Suler, 2002).

CONTROL AND EMPOWERMENT

One of the biggest advantages of therapy by telephone is that it enables clients to take a fair measure of control of the sessions. Clients can hang up at any time if they wish to terminate a session. What happens in such instances should be addressed in the contract. For example, the agreement might be that the client will recontact when the next session is due or that the practitioner will call the client within a day to confirm that the next session will take place; in the contract there needs to be a clear statement about who will take responsibility to make further contact if the client terminates a session.

There is far less intimidation when clients are in familiar surroundings than when they have to travel to a new (and someone else's) location. The telephone creates a partnership that has some equality in it – both parties are working with many unknowns that face-to-face work would answer and both are reliant on their hearing, the voice they listen to and the words expressed to form the fundamental part of the relationship. This can make therapy by telephone more attractive to some people, including those who hold controlling, senior or authoritarian positions in their life and are used to being 'in charge' of their situation, on the surface at least, as it can make the concept of their seeking help easier to accept. Equally, those whose self-esteem is not very high may be able to feel more comfortable with a voice and their imagination than with a physical presence. Such clients may feel more able to 'test' out things, to speak without feeling they have to behave in what they perceive to be an 'acceptable' manner in front of their counsellor.

TALK, SILENCE AND VOICES

Voice tone, words and pitch of both the client and practitioner are the key tools to aiding listening and understanding. Differences in the client's speed of talking, the presence of silences, the style of talking are all important for the practitioner to note and use. Talkative clients who suddenly become silent may have reached a realization or hit upon a difficulty in expressing themselves. After working by telephone for a few months, practitioners will probably notice that their listening skills become increasingly finely tuned. The slightest trace of a pause, an inflexion of the voice, occasions when the pace of talking changes are all examples of the many ways in which the client's voice can convey something that his or her words might not. An astute or distressed client will also recognize this in his or her practitioner and it has been my experience to have a client challenge me when I paused after she mentioned something that she found difficult to say. It made me realize how much care I had to take with my voice and reactions, more so than I might have done if we were in the same room and there were other distractions such as body language.

Silences can be challenging in their nature and in the context of what the practitioner should do. A client might become silent after stating something that leaves him or her feeling vulnerable, anxious, angry, sad or blocked. The practitioner must decide whether or not to break the silence and if so, when and how. She/he needs to reflect on what had been said prior to the silence to gain a clue as to the reason for the silence and an indication of how to address it or to wait. A silence of 30 seconds on the telephone can feel much longer than it would in face-to-face work. In my own work I find myself timing a client's silence and if it is longer than a minute I will break the silence; otherwise it can turn into a 'game' of who can hold out the longest, which diverts attention away from the cause of the silence in the first place. As this implies, silence on the telephone is immediate, like so much of working by telephone, as the cause is frequently directly related to what has just been said or heard and therefore needs to be explored.

Accents can be more pronounced on the telephone and can take some getting used to. A practitioner who has a strong accent should be aware of this and make it clear to the client that it is OK to ask for something to be repeated if necessary. Similarly a client with a pronounced accent may need to be asked to repeat certain phrases. This will usually be discussed quite naturally during the assessment session.

It is important to note that the sound of a pen or pencil moving on paper and typing can often be heard by the client. Any note-taking must be agreed to during the contract negotiations. It is important for practitioners to explain that they plan to make one-word notes or jot down phrases during the session and add some more detail and thoughts to them afterwards, if this is what they intend to do. Similarly, a client will detect if the practitioner is eating, drinking or smoking.

GROUP WORK

Group work by telephone is an excellent medium for short-term process work. It is of particular benefit for clients who may be unable to visit a practitioner and join a group, for example those who have a specific illness or disability, are from a particular cultural background or are isolated by their geographical location. Rosenfield and Smillie (1998) document one such telephone group and note the bonds that formed quickly between the participants, who revealed things to the others that they had not discussed elsewhere after just three (out of four) sessions.

Essentially, the practitioner needs to have access to the appropriate equipment to enable clients in their own location to join. At the appointed time, the practitioner initiates calls to all participants, one at a time, until the whole group is present. The fees that the clients pay will generally include the cost of the call in this case. The maximum practicable number for such a group is six, as any more than that makes it hard for everyone (including the practitioner) to learn all the voices and get adequate time and space. Facilitating such sessions often requires a practitioner to be proactive in the early sessions, inviting participants to contribute something until they gain confidence in speaking in the group, and it does not take more than a session or two for everyone to work out how to interrupt, how to join in. Often the clients will then ask the other group members to respond, so the practitioner's role changes to that of the person who summarizes and intervenes to push the group or an individual to look further at an issue and keep time.

WHEN THERAPY BY TELEPHONE IS NOT SUITABLE

Working by telephone does not suit everyone, practitioners and clients alike. For some, the expectation that counselling must be in a more traditional setting is important. Others may feel that they are less articulate on the telephone or that they have a strong accent that might make it hard to be understood. People who have a hearing impairment may find the telephone too difficult to use therapeutically, as may those who cannot hold a handset for a long time, although there are aids to assist with such problems.

Other factors may also place restrictions on the suitability of telephone work. For example, clients must be able to find a quiet, private room in which they will be undisturbed during their sessions. Occasionally clients who wish to have sessions during their working day, such as in their office, find difficulty in remaining undisturbed for the whole time. It is important to consider such matters during the contracting process and find ways of avoiding the problems that they can cause.

CASE STUDY ONE

Assessment

Narelle is a client who finds out about the counselling from her massage therapist. She is intrigued by the concept of never having to meet face to face and spends much of the assessment session referring to how 'ideal' for her lifestyle it is not to have to travel to meet me. She is an actor and is on the road with a touring company when we first talk. Narelle is 32 and presents with 'relationship issues' with her partner. She says she hopes that I am comfortable talking about a lesbian relationship before explaining that Simone, her partner, is also an actor, but is not with the same company and is presently at their shared, rented home.

During the assessment session we agree to talk once a week for the following six weeks in the morning, although she is quick to point out that she 'is not good in the mornings' but as she is performing six nights a week and occasional matinees, 'it has to be mornings'. We agree that sessions will take place at 10.30am for 45 minutes wherever she is for the next six weeks. Because of time differences across Australia, this means that for me some sessions will be in the afternoon.

Apart from establishing the facts we need for the contract, I leave Narelle to do the talking and I make one-word notes of what seem to me to be the keywords she uses. Afterwards I write a more detailed account of the session. She tells me that she and Simone have been involved for about seven years 'on and off', living together for the past four years, but the last two years have not been 'good years' for them. She sometimes wonders if she is too acquiescent to Simone and occasionally 'stands up to her' which causes rows and 'much grief'.

First session

Narelle calls exactly on time and sounds quite excited. She has been anticipating the session and is curious to know how things will 'turn out'. She even got up early to ensure that she had breakfast and was dressed – she wondered if she could take seriously a session held if she was in her pyjamas, even though I would not have any idea she was not dressed. I reflected to her the importance she seemed to be placing on the session and she plunged into a long explanation of how she and Simone had been talking since the assessment session and how Simone thinks the idea of counselling is ridiculous unless they are both present but how pleased she is that Narelle is doing this of her own choosing and by herself.

We explore more of their relationship and it seems that Narelle is very keen to prove to Simone something of her independence and ability to 'do this [counselling] by myself, for myself'. I ask her to tell me more about her feelings of wanting to do things for herself and there follows a torrent of examples of when she feels she has let Simone make decisions for them that she is not really happy about. Narelle's voice remains excited throughout this and she talks fast, barely pausing for breath. After some 15 minutes she slows down and I reflect on some of the things she has said, trying to explore more about the nature of their relationship.

For a few minutes Narelle answers, but with less energy than before. I point this out and she says she is 'feeling a bit strange for talking so much' and that revealing all she has to me has left her feeling exposed. We spend some time talking of how she might allow other people's perceptions to dominate her thoughts and that she often has feelings of 'inadequacy'. She then counters this with 'but my reviews have been really good' and with more exploration she identifies that it is only when she is well received on stage she feels like she is a 'whole woman', a term which she defines as feeling fulfilled in herself.

Session Two

Narelle seems quieter, although on time and enthusiastic. She is tired and is not sure what she wants to talk about. She tells me that she felt she 'talked a great deal and got a load off my chest' last time but was concerned that I might think she was not a good person. We then explored what it means to her to be a good person' and why she felt she needs approval from people around her.

This led to her talking about the sexual relationship with Simone, which she describes as 'comfortable'. We also discuss the lack of male role models in her life, although she feels that her brother was a good role model as she was growing up. He has also gone on to do 'the more traditional things – marry, have kids, buy a home and get a good job'. This brings us back to feelings of inadequacy on her part, not being good enough or doing what was expected. Her voice is expressive throughout and I am suddenly reminded that I am working with an actor. I find myself wondering if aspects of this session have been a performance, since I did not feel this last session.

Session Three

Narelle spends most of this session talking about her work and complaining about the accommodation she is now living in and how she is missing home. Her voice tone concurs with her words and when she starts to sound angry it is easy to notice this and explore the emotion with her.

I am aware of feeling that Narelle's acting skills might be more in evidence than her 'real' self from the way she uses her voice during this session. There are some pauses and phrases that I feel are being inserted deliberately to create some effect and I mention this with examples and ask her how she feels; what would Narelle the actor be saying compared with Narelle the woman on the telephone? This causes the longest silence we had throughout our sessions. I sensed it was an angry silence. After just more than two minutes, which is longer than I would usually wait, I ask her what the silence is about. She says she feels as though she's 'been sprung' and is not pleased that I *seem* to be telling her she was performing in our sessions. She acknowledges that she is 'feeling irritable' and then changes the subject back to her feeling fed up with life on the road. I am not sure why I left the silence longer than usual but it felt right to do so. The session ended with Narelle sounding flat and even weary.

Session Four

During the session we explored more about Narelle's feelings of low self-esteem and patterns of behaviour that might trigger this. At times I felt the session was getting stuck and suggested this might be a reflection of her experiences elsewhere but Narelle rejected the notion of feeling stuck until we were almost at the end of the session.

In all my sessions I inform the client when there are five minutes left and I always end on time. When I did this, Narelle suddenly rushed into what I felt was an appeasing mode, telling me she was stuck and I was right. I told her that it sounded to me as if she felt the need to leave the session with me being positive about her and our sessions. She tried to prolong the session and I mentioned this, saying that it had to end and we'd talk again next week. Afterwards I became aware that I had not felt she was performing during this session. I did not feel she was performing at any point from this session onwards.

Session Five

I introduced the session with a reminder of this being the penultimate session before our review. Narelle reacted by saying she wondered if I wanted to get rid of her and then immediately apologized, saying she knew this was my role and that of course she had remembered it was almost the end of our planned sessions.

She said she and Simone had had a row and she was thinking of ending their relationship but she wasn't sure if she could do it. That soon turned into the fact that she wasn't sure if she *wanted* to do it. We explored Narelle's past relationships, their positive aspects and their endings and touched on the impact of ending these sessions.

Session Six

Narelle had been doing plenty of work between the sessions keeping a diary of her feelings, particularly before and after talking to Simone and she told me some of what she felt were the key points that had occurred to her (the diary was not something we had negotiated as part of the contract but she told me she wrote one at various times of her life when feeling stressed). She talked of not wanting to lose the sessions but also of feeling that she needed to see if she could make things change at home by herself. She was very keen to know if she could have more sessions in the future if she wanted to and explored whether or not Simone could attend as well on their extension telephone. I said we could work together in the future and that could be as a two or a threesome.

Narelle's voice was calm and controlled. She sounded like she was confident she could work things out. She was noticeably pleased when I pointed this out. I have not heard from her again, except for a card received about six weeks after the last session quoting a passage from Gibran's *The Prophet* (1988, p. 10):

> Then said Almitra, Speak to us of Love...
> ...When love beckons to you, follow him,
> Though his ways are hard and steep...
> ...And when he speaks to you believe in him,
> Though his voice may shatter your dreams
> as the north wind lays waste the garden...

with a postscript saying:

> this is how I think of Simone and myself. We are talking *properly* for the first time in ages and I don't know where it will end but at least I now feel *more equal* in the relationship. Thanks!

CASE STUDY TWO

Dorothea is a 40-year-old mother of three, girls aged 20 and 16 and a boy of 10 who has severe learning difficulties and attends a special day school. She found out about counselling by telephone from her literacy tutor, having had tuition at home once a week from an adult literacy charity.

Assessment session

Dorothea was born in Australia of Italian parents and spoke only Italian at home. She married an Italian man when she was 18 and finally approached the literacy charity for help after her husband, 12 years her senior, had a 'heart scare'. This made Dorothea realize that he might die before her and she became determined to be able to manage on her own and care for her son. Her formal education had been very limited and she had the reading age of an 8 year old. She also could not count change and whenever she went shopping her husband would give her the money and she would hand it over not knowing if what she got back as change was correct. The reason for seeking counselling was to explore her relationship with her husband, which she described as 'hard'.

I found it difficult to understand Dorothea at first. Her accent was very strong and her speech hesitant and unclear. The words she used were simple and she repeated herself often. She thought the counselling was 'some time to talk about me and only me' and she arranged to pay for her sessions by postal (money) order as we did not meet and her husband would not write her a cheque. She told him she was using the money for an exercise class. I asked her if she could understand me fully and she said she could, but I suspected that there were times when she had not fully understood and was instead trying to work out what I meant.

We arranged four sessions in total, timed to be the day after her literacy tutoring and she used the tutoring as the way into the session each time.

Week One

This session began with Dorothea telling me how she had managed to read something and the tutor was pleased. She said she was pleased with herself too. Then she talked of all the things she was not pleased about in her life.

It soon became apparent that Dorothea had some notion that counselling was like 'magic' and that somehow I would 'make miracles happen for her'. She explained that 'being Catholic I believe in miracles and talk to my figure of Mary' in her bedroom and she thought that talking to me was a bit like that, only I talk back to her. She talked at length about her husband, his bad temper and how she often had to be in the middle between him and the eldest daughter. She told me she was not scared of him and could shout back just as loud. A few minutes later she told me he 'is not a bad man really, just like any other man' and she did not want me 'to think bad things about him'. I was aware of being very careful in my choice of words, trying to establish what she could understand. By the end of the session, after I had prompted her when I sensed she had not fully understood me, she seemed to feel confident enough in herself and also trusting enough in our relationship to ask me to explain further when necessary.

Week Two

Dorothea told me that she had not done her literacy homework because she had been too busy and how she felt about that. She spoke at length of how her husband made her angry and how much she does for everyone in the house and no one helps her. Her voice was agitated and she spoke far less clearly than when she was calm. I had to ask her to repeat herself a few times and explained that I could hear she was angry and that it would be easier for me to understand her if she could slow down a little, which she tried to do. She also told me how her son is difficult to look after and how he just watches television and only the second daughter helps her with him. Her husband hardly spoke to the boy because 'he is not like a real son'.

Whenever I sought to probe a little more into something, she paused and then carried on with what she had been saying. I wondered whether we would be able to make much progress in the four sessions.

Week Three

This week was initially focused on Dorothea's wish to lose some weight and how hard it is for her to diet and go to exercise classes. It transpired that she had been watching something on television about being overweight and how a special exercise class had made a big difference to the people on the programme. Then she talked at length of her husband and how she felt about him and their life, much of which she had said previously. He had been asking her how good the classes were since she was not looking any slimmer. I asked her if she could tell him she was using the money to talk to me and she was very definite that she would not do that.

She acknowledged that she was feeling frustrated, which was how she sounded to me. I reminded her that the next session would be our last and she said she would miss our chats.

Week Four

Dorothea's tutor had brought along play money and they had counted several dollars together although she was not sure if she could do that on her own. She returned to talking of her marriage and its difficulties. She did come up with a few ideas of how she could start off a conversation with her husband about her wish to change things a bit and had some clear examples of what she could change. I felt that counselling was unlikely to help her to consider significant changes in her relationship, or explore herself more than she had already done. At the end of the session, Dorothea said 'thank you for chatting with me. I have enjoyed talking to you and I hope we can do it again later on.'

My sessions with Dorothea made me aware of how much we expect of our clients and how articulate they need to be to facilitate building the relationship. On the other hand, it also struck me how trusting Dorothea was in the manner in which she let her feelings come out on the telephone and how she sought affirmation from me as a child might. Perhaps the telephone is not ideal for less articulate people but at the end of the four sessions Dorothea was making an action plan of how to talk to her husband.

She never implemented those specifics of which we talked, but she did call me some months later to say that he now trusted her to go out alone and to shop by herself, which was quite a big step forward for her independence and no doubt a measure of the success of her drive towards self-help through the literacy lessons and the counselling.

CONCLUSION

Telephone therapy can be an effective tool for many clients, especially for shorter term work, although there is no reason why it cannot be used over many months. The medium presents challenges to practitioners in requiring excellent listening skills and no visuals to back these up, and it does not suit everyone. There are many advantages of using the telephone for therapy including its near universal availability and the fact that it gives clients a wide choice of practitioners. People talk more freely when they feel they are not being judged and feel safe and this happens quite early on in the telephone counselling relationship, thus it may take fewer sessions to achieve the goals of the therapy than would face-to-face work.

REFERENCES

Gibran, K. (1988) *The Prophet*, Heinemann, London.

Rosenfield, M. (1997) *Counselling by Telephone*. Sage, London.

Rosenfield, M. and Smillie, E. (1998) 'Group Counselling by Telephone', *British Journal of Guidance and Counselling*, **26** (1): 11–19.

Suler, J. (2002) 'The online disinhibition effect' [Online] Available: http://www.rider.edu/users/suler/psycyber/disinhibit.html.

TBF (The Teacher Support Network) (2001), *Annual report 2000–2001*, Teachers' Benevolent Fund, London.

TISCA (Telephone, Information, Support and Counselling Association of Australia) (2000) Information Sheet. TSICA, Sydney.

6 Video counselling and psychotherapy in practice

SUSAN SIMPSON

THE DEVELOPMENT OF VIDEOCONFERENCING IN TELEHEALTH AND PSYCHOTHERAPY

The use of videoconferencing in the provision of psychotherapy is a relatively new one, although it dates back to before developments in email and Internet chat. Historically, however, videoconferencing has been used for commercial, educational and medicinal purposes over the past 40 years or so. Early developments in videoconferencing grew out of a history of rapid growth in electronic forms of communication, which were originally analogue, and most recently digital. In the initial stages, the growth of videoconferencing communications was driven by commercial organizations, such as NASA (National Aeronautics and Space Administration) in the USA. It only became used for telemedicine when individual practitioners began to utilize commercial-based equipment that was already available.

The advent of the television facilitated the development of videoconferencing and in the latter part of the 1950s medical professionals began to use closed circuit television and video communications for clinical consultations (Wootton and Craig, 1999). The Nebraska Psychiatric Institute began using videoconferencing for medical consultations, education and training in 1964. The first published study into the use of videoconferencing for psychotherapy was initiated at this same site, and used two-way, closed-circuit television for group psychotherapy (Wittson and Benschoter, 1972). Since this time, there has been a steady growth of small-scale studies and evaluations, made possible through the expansion of real-time videoconferencing and the increased availability of low-cost PCs making facilities more widely accessible. Digital communications have also improved markedly, allowing for higher quality transmission of audio and visual data.

Although the greatest area of growth and research into videoconferencing has taken place in industrial countries such as Australia, the USA and Scandinavia, with the advent of mobile video communications and satellite transmission it is becoming increasingly available to developing countries. These developments have facilitated the exchange of communications, education and training to take place across previously impenetrable boundaries.

THE CURRENT STATE OF VIDEO THERAPY – RECENT FINDINGS

Of the small number of studies that have explored the use of videoconferencing for therapy, most have been qualitative, usually case studies that have focused on user satisfaction and feasibility, or studies which have described video therapy with some face-to-face contact rather than pure video therapy. Some studies have focused on psychiatric consultations, but few have examined whether videoconferencing is suitable for the specific requirements of therapy.

Not surprisingly, due to geographical factors, the initial studies examining videoconferencing for therapy have mostly taken place in the USA, Australia and Scotland. These have tested a variety of models and therapeutic methods, including psychoanalysis (Kaplan, 1997), cognitive-behavioural therapy (Bakke et al., 2001), family therapy (Freir et al., 1999) and video-hypnosis (Simpson et al., 2002). In addition, a range of client groups have been involved, including children and families (Freir et al., 1999), adults (Simpson et al., 2001) and the elderly (Bloom, 1996). To date, there has been little investigation into whether certain client groups benefit more or less from videoconferencing-based therapies. However, so far, it has been found to hold potential for clients with a range of problems, including bulimia nervosa (Bakke et al., 2001), obesity (Harvey-Berino, 1998) and schizophrenia (Zarate et al., 1997). This chapter includes a comprehensive review of the literature into the specific field of the use of videoconferencing within counselling and psychotherapy.

SETTING UP – TECHNICAL EQUIPMENT AND TECHNOLOGY

Although a range of systems exist, most services use purpose-built videoconferencing systems initially designed for business use. The lowest cost videoconferencing equipment currently consists of a basic desktop PC with a camera mounted on top. At the other end of the spectrum are self-contained roll-about models which consist of a large (for example 29 inches) screen/video monitor on top of a coding/decoding unit with an internal microphone, tuning unit and speaker and an adjustable camera positioned directly above the screen.

A handheld remote control is used to dial up videoconferencing systems at other sites and manipulate the camera view (the view that is transmitted to the screen at the far site). If the systems are compatible, it can also be used to change the angle and magnification of the view of the far site. In most cases, the upper body of the other person is visible at both sites. Each site has the choice as to whether or not they want to see themselves with a 'picture-in-picture' facility, but most users seem to prefer not to, as it can be distracting. Audio and visual data is compressed and passed at high speed through the telecommunication network along ISDN lines (integrated services digital network) or fibreoptic cables. As the bandwidth increases (the amount of information carried by the communication lines), so does the quality of the video images in real time and, usually, the call costs. A number of recommendations are made toward the end of this chapter regarding minimizing difficulties in the use of such equipment.

VIDEOCONFERENCING VERSUS TELEPHONE THERAPY VERSUS FACE-TO-FACE: WHO PREFERS WHAT?

The descriptive data available from studies that have compared the different combinations of technology (videoconferencing, telephone and face-to-face) with therapy have not consistently found that any one approach is superior to the others. In fact, the research seems to point more to individual preferences associated with personality traits, and issues such as control, perception of personal space/distance and idiosyncratic needs and patterns within relationships. The ability to develop a therapeutic alliance has also been highlighted as an essential factor which may influence preferences of one mode over another, and this will be further examined here.

Clinical efficacy

A number of studies have suggested that therapy conducted via videoconferencing may be equivalent to face-to-face sessions in terms of both client satisfaction and clinical effectiveness. There is a distinct lack of research that has examined the clinical efficacy of video therapy in any depth, with no randomized controlled trials having been completed at the time of writing.

The Wittson and Benschoter (1972) study noted above used two-way, closed-circuit television for group psychotherapy and compared this with control groups who received face-to-face group psychotherapy. After six sessions, results showed that the effectiveness of therapy was also unaffected by the presence or absence of technology. Other factors, such as group composition and choice of leader, influenced outcome more than the mode of therapy delivery per se. More recently, Harvey-Berino (1998) compared face-to-face with videoconferencing group behavioural therapy for the treatment of obesity. Both conditions were found to lead to significant changes in eating and exercise patterns and were effective in leading to weight reduction. Most clients were highly satisfied with videoconferencing, even those who were aware that they could have seen a therapist face to face. Although at first over 50 per cent were hesitant about communicating via videoconferencing, with experience they became more comfortable with the technology.

Schneider (1999) compared brief cognitive-behavioural therapy via video link, two-way audio and face-to-face with a waiting list control group. Results showed that there was no significant difference between treatment groups across a range of outcome measures, but all were superior to no treatment. Schneider suggests that, in future, it will be essential to consider such factors as presenting problems, personality types and comfort levels in order to determine which clients are most suited to which mode of delivery.

A small number of descriptive case studies have also shown promising results. Bakke et al. (2001) treated two clients with bulimia nervosa via videoconferencing using a cognitive-behavioural therapy (CBT) model and reported that both were abstaining from bingeing and purging at the end of treatment and follow-up. Studies carried out by the author (Simpson, 2001; Simpson et al., 2001) also described clinical improvement in a sample of clients treated using CBT, as shown by ratings on clinically validated questionnaires, self-report and ratings by general practitioners. Simpson (2002) found that a group of eleven clients felt significantly more confident to deal with their problems (including insomnia, flight phobia, social anxiety, binge eating) following a single session of video-hypnosis.

Kaplan (1997) described two psychoanalytic case studies, one of whom was seen almost exclusively via videophone, and the other as a combination of face-to-face, telephone and videophone sessions. Videophone sessions were set up so that the analyst could see the client lying on the couch from the same angle as he would have had in his office. His view was that the videophone did not produce any subjective difference to the client's or his own experience of therapy, and when he examined his progress notes post-therapy, he was not able to distinguish which sessions had taken place with or without the technology. He humorously noted one benefit of videophone sessions by describing an incident whereby after accidentally sleeping late one morning, his client was able to avoid attending the session late by the fact that he could switch on the videophone in his home wearing his pyjamas.

Client satisfaction

In some cases, clients have expressed a distinct preference for video therapy over face-to-face sessions. However, most have not had any choice over the mode of their therapy, so it is not known whether this represents general satisfaction with the treatment they received or a more specific preference for videoconferencing. Simpson et al. (2001) found that 9 out of 10 clients offered therapy via videoconferencing expressed satisfaction with the service, and some preferred it to face-to-face contact. Clients' comments indicated that they experienced video therapy as less embarrassing and confrontational than face-to-face contact, and one said 'it was easier to talk to someone on a screen than having someone invading my space'. Some clients felt that they were more easily able to express difficult feelings via videoconferencing and that the extra distance made them feel safer. Similar results were found by Bakke et al. (2001) who treated two women with bulimia nervosa via videoconferencing, using a manual-based cognitive-behavioural model. Results showed that clients valued the privacy and anonymity of video therapy, and they commented that it was less intimidating than face-to-face sessions. They also valued the fact that they were not required to travel to sessions.

In another recent study (Simpson et al., 2002), 11 clients attended a single session of hypnosis via videoconferencing. One-third of clients preferred video-hypnosis to the prospect of face-to-face hypnosis and one-third had no preference. Even those who expressed a preference for face-to-face hypnosis indicated that they would like further video-hypnosis sessions. A number of reasons were given for preferring video-hypnosis over face-to-face hypnosis, including a greater sense of control, an increased sense of being alone in the room and therefore less under scrutiny and a consequent reduced sense of self-consciousness. On the other hand, the views of those clients who indicated a preference for face-to-face sessions suggested that videoconferencing can lead to a reduced social presence and sense of connection with the therapist.

Thus, not all clients may be suited to video therapy. One client who did not experience video therapy in such a positive light reported that he feared that his sessions might be being recorded or watched by others and that this information may be used against him in the future. He described the video therapy as 'dehumanising' and 'unsettling', and felt that he would need time to build up a therapeutic relationship in a face-to-face setting in order to benefit from therapy. It was noted that this client's longstanding avoidant and anxious personality traits, and in particular his difficulty trusting others, may have made it more difficult for him to benefit from therapy in this context (Simpson et al., 2001).

Another client found that she missed the intimacy of face-to-face sessions, and often compensated by sitting so close to the screen that only a magnified version of her nose was visible on the therapist's screen. Although she was frequently asked to sit back a little, she habitually returned to her former position. She also compensated in the form of telephoning the therapist and sending her long letters between sessions. Although this client may have found the closeness of face-to-face sessions more acceptable, she was in fact extremely positive in her ratings of video therapy sessions, and in retrospect it seemed that these may have in fact facilitated the development of a more independent and self-contained aspect of her personality. It may be then that for some clients with issues associated with dependency and lack of sense of self, video therapy may provide additional therapeutic value, even if it is not their first choice.

Simpson (2001) suggested that there was a trend for those clients with more complex problems to rate lower levels of satisfaction with video therapy than those with shorter term, simpler problems. This was also found by Ghosh et al. (1997), who suggested that the self-consciousness and awkwardness reported by a client whilst discussing sensitive issues were more attributable to her general character and relationship difficulties than to the technology. This suggests that these difficulties may also have arisen if therapy had been conducted on a face-to-face basis.

A number of possible factors may influence clients' preferences, including their level of previous experience with videoconferencing and the particular nature of their difficulties. For example, those who feel that they lack control in their lives and relationships in general might prefer the extra control offered by videoconferencing. Similarly, those who feel intimidated or self-conscious about talking about themselves or their problems may also prefer the distance and control offered by videoconferencing. A sense of control has often been cited as a major factor influencing clients' level of satisfaction with video therapy. Allen et al. (1996) found that in video therapy clients often perceive a greater sense of control than with face-to-face therapy, due to being able to move out of the camera view or even the room and being able to switch off the equipment if they so desired. According to Omodei and McClennan (1998), this sense of control can be further heightened by giving clients the remote control panel so that they have the opportunity to manipulate sound and picture. Those clients who need a greater sense of control within their sessions may find this particularly appealing, but those who are particularly fearful of using technology for communication may find it confusing or anxiety-provoking.

The small number of studies that have compared video therapy with telephone therapy have been inconclusive in their findings. Schneider (1999) compared brief therapy across video, two-way audio and face-to-face modes, and suggested that although individuals may be more suited to one mode over another, most adapt to any given mode with time and experience. Although he found that drop-out rates were higher for the two-way audio and video modes than for face-to-face, he suggested that a longer term therapy may have produced different results by giving clients time to adjust. In comparing video, audio and face-to-face modes of therapy, Kaplan (1997) found videophone to be a more acceptable means of therapeutic contact than the telephone, due to being able to see the other person's facial expressions and kinaesthetic cues, which enhanced both conversational synchrony and understanding.

Therapist satisfaction

Few studies have examined therapist preference for one mode of communication over another. It will be of utmost importance to identify those factors that influence whether or not therapists choose to engage in one mode over another if new forms of technology are to be utilized in the provision of therapy in the future. Those studies which have explored this area suggest that as with clients, individual therapists differ in their attitudes to video therapy. Although therapists often tend to be more cautious about video therapy than clients, this initial reluctance tends to recede with experience and practice (Nagel and Yellowlees, 1995). Therapists are often relieved to learn that they can continue to use their usual therapy techniques and strategies, and only need to learn a few minor extra skills, such as the process of turn-taking, in order to become effective video therapists (Omodei and McClennan, 1998).

Simpson et al. (2001) reported a high level of therapist satisfaction with videoconferencing sessions, and noted that although some initial adjustments were required in communication style and pacing, adaptation was relatively quick. McLaren et al. (1996) reported that although some therapists claimed to experience higher levels of fatigue following video therapy sessions, others remarked that they felt more relaxed, and were even able to take off their shoes and put their feet on a stool out of the view of the camera.

A recent survey carried out in Aberdeen, Scotland, looked at reasons why certain users are more or less likely to utilize videoconferencing facilities for clinical purposes (Mitchell et al., 2003). Therapists from the psychiatric hospital along with medical doctors from the local A&E department were asked to fill out brief questionnaires and undertake a short interview to elicit their views. Results suggested that people were more confident in the use of videoconferencing for clinical purposes when they had undergone hands-on training in the use of videoconferencing equipment, followed by opportunities to use it regularly thereafter, with the availability of top-up training when required. It was also suggested that self-rating of personality factors such as risk-taking behaviour and openness to new experience may be positively correlated with willingness to use videoconferencing for clinical purposes.

At present, most technical training available is through videoconferencing manufacturers (for example Sony), who can offer this as part of the deal when purchasing equipment, or in-house training by experienced users who are already familiar with the equipment. There is currently little formal training available on the use of videoconferencing for therapeutic purposes, although this may at present take place informally at local sites where video therapy is offered. It would seem essential that videoconferencing training programmes be made more available to ensure that the highest possible standards of care are upheld, and so that users will have the skills and confidence to offer these services in the future.

VIDEOCONFERENCING AND THERAPEUTIC ALLIANCE

The development of a positive therapeutic rapport is a well-documented prerequisite for making insights and changes in psychotherapy. A small number of studies have examined the way in which the 'videoconferencing environment' interacts with certain factors such as empathy and social presence, thereby playing an influential role in the establishment of the therapeutic relationship.

Fussell and Benimoff (1995) reported that the establishment of an alliance is dependent on the availability of non-verbal cues such as eye gaze and gestures. Eye gaze is purportedly significant in helping communicators to establish speaking turns and recognize whether or not the other is engaging in the discussion. Gestures may further contribute to conversation, by facilitating speaking turns and providing cues which clarify what is said. It is also suggested that gestures may facilitate verbal fluency and message formulation, and that conversation may become more disjointed when hands are restricted. This may be particularly pertinent for client groups such as the Australian Aborigines whose communication systems include a range of hand gestures whilst speaking (Hodges, 1996).

Reservations have been expressed about whether it is feasible to develop a positive therapeutic rapport via videoconferencing. When using low-quality videoconferencing, communication can be compromised by sound delays, lack of lip-voice synchronization and poor image quality, which can inhibit appropriate turn-taking and interpretation of facial expressions, and thereby detract from rapport-building in the initial stages of therapy (Kirkwood, 1998). Sound delays can disrupt the normal rhythm of speech. For example, when therapists use minimal prompts such as 'uh huh' and 'hmmm' to facilitate the other person's speech, this may be picked up at the far site at slightly the 'wrong' point in the flow of conversation, thus causing confusion. In addition, hand and arm gestures can become pixelated (the screen image becoming interrupted by jerkiness or the picture breaking up) due to the delay in image transmission, thus reducing the quality of this aspect of non-verbal communication. Direct eye contact is also often not possible with videoconferencing, as the camera is positioned above the viewing screen. However, a number of studies (Ghosh et al., 1997; Simpson, 2001) have found that both therapists and clients adjust to these differences, and rate the therapeutic alliance highly in spite of them. Similarly, Simpson et al. (2001) suggest that empathy and warmth can be transmitted via videoconferencing, and that although traditional methods (such as handing clients a box of tissues) may not be feasible, alternatives can be developed through facial expression, voice tone and the respectful use of silences. As the technology develops, picture quality improves and delays in sound are diminished, the precision and clarity of communication will become greater. It may be informative in the future to evaluate whether there is a relationship between the quality of communication and corresponding levels of therapeutic alliance of video therapy sessions.

Some have expressed concern that therapeutic rapport may be compromised by reduced social presence and less spontaneity during videoconferencing interactions (Allen and Hayes, 1994). However, in contrast, Capner (2000) found in her review of the literature that clients have largely been satisfied with the social presence in video therapy, and in fact some have commented that they became so used to it that they forgot they were not in the same room as their therapist. Schneider (1999) suggested that most therapists are able to use the cues available to them whether they be audio or visual and find ways to develop a positive alliance regardless of media mode.

The current author has found that the use of peripheral forms of communication such as email, letter writing and fax can further support the therapeutic alliance, particularly for those clients who require more intense input. Telephone support has been offered occasionally, both as a substitute for therapy sessions when technical breakdown has prevented videoconferencing sessions from taking place and as an adjunct at times of crisis.

THERAPEUTIC PROCESS ISSUES

Although the process of therapy tends to be similar for video therapy as in face-to-face settings, it is prudent to consider the ways in which the presence of technology may influence the different elements of therapy.

When considering the initial aspects of establishing a therapeutic relationship, there is some variation in the suggestions made by different authors. Gammon et al. (1998) suggest that in psychotherapy supervision, the establishment of a face-to-face rapport over several sessions prior to engaging in videoconferencing sessions contributes to the mutual trust and respect which is required for such a relationship to function effectively. However, Simpson et al. (2001) and Kaplan (1997) reported that it is feasible to develop positive therapist–client rapport after just one initial face-to-face assessment session. There is a lack of evidence to date examining whether it is possible to develop an adequate therapeutic relationship solely through videoconferencing sessions, as most studies have used a combination of videoconferencing and face-to-face sessions. This author is involved in such a study, examining both therapeutic alliance and clinical effectiveness of therapy conducted solely via video link for clients with bulimia nervosa.

It is recommended that therapists remain aware of certain boundary issues that may be particularly relevant to working by videoconferencing (for example dealing with clients taking snacks and drinks into sessions). In addition, it is important to maintain the normal boundaries which are utilized in face-to-face work wherever possible, to ensure that both therapists and clients alike maintain an awareness of what is expected of them in order for the therapy to proceed effectively.

RECOMMENDATIONS

Clinical

- Therapists with 'at risk' clients should remain in close contact with the client's GP (family doctor) or other relevant health professionals; and ensure that adequate support is available to clients between sessions and in case of a crisis.

- Therapists should allow time to adjust to using the videoconferencing system, and note that in general any initial discomfort is likely to dissipate with experience.
- Pre-therapy client information sheets should include information about the therapist, basic details on how video therapy works and how to access the service.
- Provide information sheets about procedures in the case of technical failure. This should involve the therapist re-establishing the videoconference call as soon as possible, or at the least by making contact by telephone. Telephones should be available for use in all videoconferencing rooms. When sound is lost, or other technical problems occur, these keep both the client and technicians informed (Goss, 2000).
- Speakers need to learn the etiquette of turn-taking in conversation, such that each person ensures that the other has finished speaking before starting, rather than relying on lip movements which may be out of synchrony with sound (Cukor and Baer, 1994). Clients are most comfortable when therapists use their normal style of communication – attempts to compensate for the extra distance in video therapy can be perceived as artificial (Capner, 2000).

Technical

- Quality of calls are determined by picture size, response delay and frame rate. With lower quality systems, users should keep fast movements or gestures to a minimum. As the quality of calls increase, facial expressions and signs of distress and tearfulness become significantly easier to detect.
- A solid colour (especially dark) background is best for maximizing picture quality and a dark heavy curtain can achieve this effect and simultaneously improve the room's soundproofing.
- Staff should be trained in operating the videoconferencing system, including focusing the camera, using associated equipment, timing speech and turn-taking. In addition, users may benefit from practising staying within the camera view, and in giving feedback to the person at the far site about volume and positioning.
- Block booking reduces stress associated with the complex logistics encountered with booking videoconferencing systems and per-

sonnel to operate them at both sites (when needed), and also provides consistency for clients.

- In order to obtain the best possible balance between capturing body movements and eye contact and facial expressions in the picture, the camera should be positioned to the point where head, shoulders and arms are contained within the image (Fussell and Benimoff, 1995).
- Lighting should be bright enough in order to produce a clear image for projection to the far site. In the case of video-hypnosis, a balance needs to be reached between keeping lighting dim to enhance client comfort, whilst ensuring that client breathing patterns, facial expressions and so on remain visible.

CASE STUDY

Presenting problem

Peggy presented for therapy with long-term problems associated with her weight and eating, as well as recent health problems including gallstones and chronic pain. Her weight at the time of referral was 120.7 kg, which gave her a body mass index of over 45 (normal range = 19–25). In order to safely operate on her gallstones she had been advised that she needed to lose at least 25 kg. In addition, over the past year her weight and size had exacerbated her chronic pain, and the combination of these two factors made physical exercise difficult.

At the point of seeking help, she lived on a remote island off the south coast of England. She described feeling alone, isolated and in a great deal of pain. Her problems made travel and exercise difficult, and she was therefore largely housebound.

Formulation

A full assessment of Peggy's eating patterns confirmed a diagnosis of 'overeating associated with other psychological disturbances' (ICD-10 diagnostic system). In addition, her chronic pain, which was caused primarily by gallstones, was found to be exacerbated by stress associated with physical exertion and lack of physical fitness, overeating episodes and boredom.

Peggy was also experiencing clinically significant levels of anxiety and depression. She believed that she was socially undesirable and unattractive, needing to prove her worth through losing weight, and striving for perfection. Her weight and shape became a way of measuring her self-worth. Her desire to be more physically attractive led to an attempt to 'overcontrol' her eating, through restrictive dieting. Paradoxically, through focusing excessively on her eating and experiencing sensations of starvation after skipping meals, she began to compensate through periods of overeating and binge-eating, whereby she would lose control and eat large quantities of the foods that she had previously forbidden herself.

Treatment

Treatment took place over 12 sessions with a clinical psychologist and 8 sessions with a dietician, with follow-up sessions planned on an intermittent basis. Peggy attended her sessions at the local mental health department on the island, where a PC-based teleconferencing system was assembled with a camera on top. The clinical psychologist and dietician used a teleconferencing system, based on site at the hospital in Aberdeen. The client and therapists could see the other person on their screen from mid-torso.

Sessions involved a combination of direct therapy work using a cognitive-behavioural therapy (CBT) approach, and hypnosis sessions. The CBT sessions focused on enabling Peggy to recognize and challenge unhelpful thinking patterns and build her self-esteem through learning to value and accept herself. She was helped to re-evaluate her sense of worth and attractiveness by means other than focusing exclusively on her weight and shape. Her eating difficulties were addressed initially through developing her awareness of patterns by keeping food diaries (of all food/liquid consumed, and associated thoughts/feelings) which were emailed to therapists on a weekly basis for discussion in sessions. Letters were written and emailed to Peggy by the therapist throughout the course of therapy to reinforce the content of sessions.

Hypnosis sessions focused on assisting Peggy in managing her chronic pain through suggestions about the creation and transfer of symptoms of analgesia. It was anticipated that learning to use self-hypnosis would increase her sense of control and efficacy over her pain, and that she would then be less likely to turn to eating as a way of dissociating from this experience.

Peggy was easily hypnotized, and in fact became deeply relaxed in a relatively short time period. Arm levitation was particularly effective as a deepening technique. Her arm could be seen moving upwards on the screen, and signs that she was becoming relaxed (deeper breathing, diminution of body tension) were easily detectable via the videoconference image. Although the sessions were conducted at a low bandwidth, the image of the client was quite still during hypnosis, so the quality of the picture remained relatively good. A large, supportive chair was provided to maximize the client's comfort during sessions, and the camera was adjusted accordingly to focus on her.

Due to the delay caused by using a low bandwidth, some adjustment was required in communication to ensure the other person had finished speaking before beginning to speak oneself (turn-taking). However, during hypnosis this caused little disruption as the client was rarely required to speak and communicated mainly through nodding at relevant junctures. Some improvisation was also required in terms of signalling within hypnosis sessions, in order to convey messages between patient and therapist at relevant junctures. For example, Peggy was asked to signal by nodding rather than raising a finger, which may not have been as easily detectable using videoconferencing.

The psychologist spent the first part of the session reassuring Peggy that if any technical problems occurred, or if the tele-link was temporarily lost, she would be able to open her eyes and be in control. She also repeated during the tele-hypnosis session the words: 'If for any reason you need to open your eyes, you'll be fully orientated and able to do what you need to manage the situation.' As Peggy's sessions took place within the mental health department in the local hospital, other health professionals were usually present in the building and the clinical psychologist conducting the sessions would have been able to telephone them if any particular clinical difficulties had arisen in the course of tele-hypnosis sessions. If any concerns had been raised about Peggy's safety, then it would also have been possible to contact the local psychiatrist or one of the general practitioners to notify them about the situation.

Outcome

Peggy made a number of significant improvements in video therapy. She lost more than 5 per cent of her bodyweight over the course of therapy through an increase in regular healthy eating and physical exercise. Results from Peggy's questionnaires and self-report also reflected a positive effect of video therapy on her self-esteem, and she was no longer clinically depressed or anxious following her course of sessions.

Qualitative feedback: At the post-therapy telephone interview, Peggy commented that hypnosis via videoconferencing was a 'totally positive experience' for her. She noted that she actually found it easier to communicate through videoconferencing as she felt it was less personal than face-to-face sessions whereby she may be watched and her body language analysed. As a result, she felt that with videoconferencing there were 'no boundaries' and she felt more able to 'say exactly what I felt at the time'. This was also helped by the fact that the therapist was conducting the therapy from outside the local community, and therefore she perceived that there would be greater confidentiality, and little risk of running into each other at community events or the local supermarket.

ETHICS AND LICENSING

Different regulations and qualifications are required in order to practice between states in the USA and Australia, and between countries in Europe and elsewhere. It may be prudent to consult a legal specialist in order to ascertain responsibility and licensing procedures. Seeking client consent which specifies that they accept that all services received from the therapist are considered to be provided entirely at the therapist's base may protect to some degree from clients who may sue on the grounds of non-residence (Seeman and Seeman, 1999).

Reimbursement for services also becomes problematic across state and international boundaries. Private insurers have not yet agreed to pay for video therapy, and the difficulties which therapists may experience in obtaining reimbursement may lead them to avoid offering these services (Capner, 2000).

Duty of care has also been highlighted as a potential area for liability, resulting from the establishment of a therapeutic relationship. Capner (2000) suggests that when therapists are working in collaboration with other clinicians at far sites, it is imperative to clarify professional boundaries at the outset for professional indemnity purposes. It may be considered unethical by some to offer therapy via videoconferencing at such an early stage, when there is a lack of controlled trials to show its efficacy. However, it may also be claimed that withholding therapy services from clients living in remote areas or those who would prefer to work in this way when the technology is readily available may be equally unethical. It is therefore also recommended that client consent be sought on the grounds that there is a lack of evidence into the efficacy, reliability and validity of video therapy.

Although concern may be expressed at the level of confidentiality of sessions, given that clinical information is being transmitted via videoconferencing, video therapy is likely to be more secure than face-to-face therapy, as there is usually less likelihood of being interrupted, and the probability of being able to break into between two and six ISDN lines is minimal due to the complexity of such an operation.

CONCLUSION

Until the present time research has been mostly qualitative and self-report in nature, focusing on clients' and therapists' experience of the acceptability and feasibility of video therapy. Despite some misgivings and the negative experiences of some, most therapists and clients generally adjust to using videoconferencing within a few sessions. Clients appear to be at ease with video therapy more quickly than therapists and some have even commented that they had forgotten that the therapist was not in the same room. Many clients have expressed a preference for video therapy due to an increased sense of control, feeling less intimidated/confronted and less self-conscious than in face-to-face settings. A minority have reported that the presence of the technology made them feel more distanced and isolated, which may be made worse by the poorer transmission quality at lower bandwidths.

Further research is needed to determine whether certain client groups are more suited to video therapy than others, depending on such factors as familiarity with the technology, age, gender, personality characteristics, presenting problems and previous therapy experiences. One might predict that the younger generation who have been brought up on text-messaging and email may be expected to feel more at ease with the technology. Larger randomized controlled trials are required to determine the relative efficacy of video therapy in comparison with face-to-face therapy, and a multi-centre approach may be needed to recruit sufficient numbers to accomplish this.

As the technology becomes more sophisticated and less costly, videoconferencing facilities will be increasingly available to those living in remote areas, both in health centres and peoples' homes. It is likely that as videoconferencing facilities become more widely accessible, the demand for video therapy services will increase correspondingly and therapists will have more opportunities to develop skills for working in this area. It is also likely to become relevant and applicable in a wider range of settings, such as prisons, large cities with heavy traffic congestion and developing countries. As they do so, therapists will be required to develop specific standards of care via videoconferencing, and negotiate ways of standardizing qualifications and/or registration with professional bodies which are recognizable across geographical and international boundaries.

ACKNOWLEDGEMENT

The author would like to acknowledge the assistance of Stephen Bell and Emma Morrow in the preparation of this chapter.

REFERENCES

Allen, A. and Hayes, J. (1994) 'Client satisfaction with telemedicine in a rural clinic', *American Journal of Public Health*, **84** (10): 1693.

Allen, A., Roman, L., Cox, R. and Cardwell, B. (1996) 'Home health visits using a cable television network: user satisfaction', *Journal of Telemedicine and Telecare*, **2**: 92–4.

Bakke, B., Mitchell, J., Wonderlicht, S. and Erikson, R. (2001) 'Administering cognitive behavioral therapy for bulimia nervosa via telemedicine in rural settings', *International Journal of Eating Disorders*, **30**(4): 454–7.

Bloom, D. (1996) 'The acceptability of telemedicine among healthcare providers and rural patients', *Telemedicine Today*, **4**(3): 5–6.

Capner, M. (2000) 'Videoconferencing in the provision of psychological services at a distance', *Journal of Telemedicine and Telecare*, **6**: 311–19.

Cukor, P. and Baer, L. (1994) 'Human factors and issues in telemedicine: a practical guide with particular attention to psychiatry', *Telemedicine Today*, **2**(2): 9–18.

Freir, V., Kirkwood, K., Peck, D., Robertson, S., Scott-Lodge, L. and Zeffert S. (1999) 'Telemedicine for clinical psychology in the Highlands of Scotland', *Journal of Telemedicine and Telecare*, **5**: 157–61.

Fussell, S.R. and Benimoff, I. (1995) 'Social and cognitive processes in interpersonal communication: implications for advanced telecommunications technologies', *Human Factors*, **37**: 228–50.

Gammon, D., Sorlie, T., Bergvik, S. and Hoifodt, T.S. (1998) 'Psychotherapy supervision conducted by videoconferencing: a qualitative study of user's experiences', *Journal of Telemedicine and Telecare*, **4**: 33–5.

Ghosh, P.G.J., McLaren, M. and Watson, J.P. (1997) 'Evaluating the alliance in video-link teletherapy', *Journal of Telemedicine and Telecare*, **3**(Suppl. 1): 33–5.

Goss, S. (2000) A trial of video therapy. Unpublished manuscript.

Harvey-Berino, J. (1998) 'Changing health behaviour via telecommunications technology: using interactive television to treat obesity', *Behaviour Therapy*, **29** 505–19.

Hodges, M. (1996) 'Online in the Outback' See http://www.mit.edu/afs/TrendOutback.html.

Kaplan, E.H. (1997) 'Telepsychotherapy – psychotherapy by telephone, videophone and computer videoconferencing', *Journal of Psychotherapy Practice and Research*, **6**: 227–37.

Kirkwood, K. (1998) 'The validity of cognitive assessments via telecommunications links', PhD Thesis, University of Edinburgh.

McLaren, P.M., Blunder, J., Lipsedge, M.L. and Summerfield A.B. (1996) 'Telepsychiatry in an inner-city community psychiatric service', *Journal of Telemedicine and Telecare*, **2**: 57–9.

Mitchell, D., Simpson, S., Ferguson, J., Smith, F. (2003) NHS staff attitudes to the use of videoconferencing in the provision of clinical services. Poster Abstract. Conference Proceedings: Telemed '03 from research to service delivery. 10th International Conference on Telemedicine and Telecare. 29–30 January 2003.

Nagel, T. and Yellowlees, P. (1995) 'Telemedicine in the top end', *Australasian Psychiatry*, **3**: 137–9.

Omodei, M. and McClennan, J. (1998) ''The more I see you?' Face-to-face, video and telephone counselling compared. A programme of research investigating the emerging technology of videophone for counselling', *Australian Journal of Psychology*, **50**(suppl.): 109.

Schneider, P.L. (1999) 'Psychotherapy using distance technology: A comparison of outcomes', American Psychological Association, Boston: Paper Presentation. http://www.outreach.uiuc.edu/~p-schne/research/technologyvsface.html.

Seeman, M.V. and Seeman, B. (1999) 'E-psychiatry – the patient–psychiatrist relationship in the electronic age', *Canadian Medical Association Journal*, **161**: 1147–9.

Simpson, S. (2001) 'The provision of a telepsychology service to Shetland: client and therapist satisfaction and the ability to develop a therapeutic alliance', *Journal of Telemedicine and Telecare*, **7**(suppl. 1): 34–6.

Simpson, S. (2002) Tele-hypnosis – The Provision of Specialised Therapeutic Treatments via Teleconferencing. Unpublished manuscript.

Simpson, S., Deans, G. and Brebner, E. (2001) 'The delivery of a tele-psychology service to Shetland', *Clinical Psychology and Psychotherapy*, **8**: 130–35.

Simpson, S., Morrow, E., Jones, M., Ferguson, J. and Brebner, E. (2002) 'Tele-hypnosis – the Provision of specialised therapeutic treatments via teleconferencing', *Journal of Telemedicine and Telecare*, **8** (suppl. 2).

Wittson, C.L. and Benschoter, R. (1972) 'Two-way television: helping the medical center reach out', *American Journal of Psychiatry*, **129**(5): 624–7.

Wootton, R. and Craig, J. (eds) (1999) *Introduction to Telemedicine*. Royal Society of Medicine Press, London.

Zarate, C.A., Weinstock, L., Cukor, P., Morabito, C., Leahy, L., Burns, C. and Baer, L. (1997) 'Applicability of telemedicine for assessing clients with schizophrenia: acceptance and reliability', *Journal of Clinical Psychiatry*, **5**(1): 22–5.

7 Video and telephone technology in supervision and supervision-in-training

PHILLIP ARMSTRONG AND H. LORI SCHNIEDERS

In 1985 the Internet connected a mere 2,000 computers, but by the turn of the century it connected 30 million. The Internet is continuing to double in size every year and it is projected that by 2005 there will be more than one billion computer users. And yet counsellors have been slow to embrace the use of computer technology in the supervisory process (Schnieders, 2000). Now, satellite and wireless systems are providing users with 'anytime, anywhere' communication (Sabella, 1999), including the use of video and telephone links. Video, in particular, as a means of providing the basis for supervision, (especially live, in-session observation), has only come to the forefront in the last five years.

The following chapter will examine the use of videoconferencing and the telephone as a means for providing supervision of counselling and psychotherapy practitioners, and although specific contexts are used for the sake of discussion, much of the following material is readily transferable to other contexts. The first section, which focuses on video-based supervision, offers a special focus on the issues in relation to counselling and therapy practitioners in training and is based in part on work undertaken in that setting. Reference has been made to live supervision as a particularly challenging and relevant situation, although much of what follows could also be applied to other types of supervisory sessions as well. The second part of the chapter looks at telephone supervision, drawing particularly on the experience of such methods in Australia where geographical barriers are a particular spur to their development and acceptance. This is not to say, however, that those in other places, whether in south-east England, downtown Los Angeles, the Philippines or anywhere else would not also have reasons for considering the same options for their convenience and other advantages over more traditional arrangements.

LIVE SUPERVISION

During the last 30 years, live supervision of practitioners-in-training has taken many forms, from telephone communication links between the supervisor and trainee, to supervisors physically entering sessions to provide consultation, to having headphones for trainees and microphones for the supervisors. There is also the 'bug-in-the-ear' method (which is valuable in other settings, such as family therapy), in which trainees wear small radio receivers in their ear and the supervisor provides on-the-spot direct consultation.

Neukrug (1991), in his article 'Computer-assisted live supervision in counsellor skills training', was among the first to examine the use of computers for live supervision. He described a process of using a one-way mirror during practice counselling sessions. The supervisor provided non-intrusive feedback to the counsellor-in-training by typing on a computer keyboard; the trainee, who was situated to view the computer monitor, was able to read the feedback as it appeared. From his work with this form of supervision, Neukrug concluded that 'the computer could potentially bring even quicker and stronger [counselling] skills acquisition'.

Live supervision differs from other forms of supervision because the supervisor observes the actual counselling session as it happens and can make immediate supervisory interventions (Montalvo, 1973; Birchler, 1975; West et al., 1989; Neukrug, 1991). To an even greater degree than that of a video camera or tape machine, the supervisor's presence can produce anxiety in supervisees and self-consciousness about being observed (Bowman, 1980; Liddle, 1986).

At a time when the explosive growth of the Internet has impacted most of the world, having a computer in one's office is, for many, an accepted element of the counselling practitioner's environment. This leads to the appealing prospect of using a webcam attached to a computer for live supervision. It also diminishes apprehension and reduces the fear of making a mistake, because the counsellor-in-training is in a familiar environment, generally surrounded by their 'own' possessions. Furthermore, with the addition of text discussion boards used in conjunction with video or audio links, it is also possible to conduct group supervision through 'virtual' discussions in real time through multiple links.

VIDEOCONFERENCING

Currently, access to most supervision is limited by the availability of qualified supervisors and the distances either supervisor or supervisee must travel to attend sessions. In training situations, supervisors must sometimes travel great distances to observe students and interns on site – or the students must travel to see the supervisor. In a time of increasing fuel prices and demands on supervisors' time, especially in university settings, it has become less cost-effective for many supervisors to make site visits where they are required (Schnieders, 2000), especially those who cover large geographical areas and are sometimes required to travel for several hours to work with one supervisee. Even where supervision is arranged such that the supervisee travels to see the supervisor, the travel costs and time will almost certainly be comparable overall, albeit spread across a number of individuals.

To minimize the costs of travelling to a site, one-on-one supervision time can be accessed by Internet connection to password-protected sites similar to iVisit (www.ivisit.com) or CU-SeeMe (www.wpine.com), or sites provided by an institution. This form of technology for desktop videoconferencing has improved vastly. It is now possible both to view and hear each other with minimal delay, unlike several years ago, when time delay caused the voice to occur five to ten seconds after the picture (see also Chapter 6). Using this format, prearranged times are set for the practitioner-in-training and the university supervisor to log on to the site. Both supervisor and student can see each the other and conduct real-time supervision sessions via split-screen videoconferencing. A distinct advantage, especially in training settings, is that such delivery systems also provide the supervisor with the means to capture the video or audio images for processing with the supervisee at a later date.

Many videoconferencing packages also provide a white board, similar to an Internet chat room environment (see Chapter 4), where typewritten messages can be shared between student and supervisor. They may also allow for group supervision, with multiple users logging on at the same time.

POTENTIAL PROBLEMS AND RISKS

The possibility exists for breach of confidentiality of client information that is transmitted electronically over computer networks during supervision (Sampson and Pyle, 1983; Hannon, 1996; Bloom, 1998). The use of public servers as a supervision resource can violate confidentiality when client data provided to facilitate the exchange of ideas about client issues and interventions are detailed enough to identify a specific client (Casey et al., 1994; Marino, 1996), as would generally be the case with video images. Additionally, unauthorized access to the conferencing site may be particularly problematic. Therefore, supervisors need to be vigilant of potential threats to confidentiality and become familiar with and consistently use appropriate security methods.

Even though equipment and connection costs have been drastically reduced in recent years, they still remain out of the financial reach of many individuals. Technical competency is also an issue. The supervisee must be able to connect the hardware necessary for online supervision. Next, he or she must be able to access the Internet, connect to the appropriate net-meeting site, log into the site and manoeuvre the mouse, webcam and microphone, making good use of the necessary software package while videoconferencing with the supervisor.

For supervision to be effective, physical privacy to facilitate the supervision process is necessary, as it is with remote therapeutic services too. Any form of supervision would be liable to be compromised if the supervisee were to be accessing the Internet in a library or computer lab.

CASE STUDY EXAMPLE AND DISCUSSION

The course from which the following example was taken had existed before it was developed for the Web. What was known about effective on-campus supervision was used for the creation of the same aspects of supervision for a new environment. This should not imply that only on-campus teaching and supervision practices were applied to the new environment. Web-based environments made new demands on the students and required the instructor/supervisor to develop a new inventory of online supervision strategies while embracing the rapid developments in technology.

Pedagogic decisions for online supervision were in a large part made during the development process, although modifications continued to be made as the process evolved. For example, the first student engaged with for video supervision was placed in an internship in a neighbouring state, approximately 450 miles from the university. The intern/supervisee was a technology novice but far from technophobic and willing to engage in the process. Equipment and its working order became the focal point of this interaction. This first supervision experience was at the beginning of webcam use in early 1999. Telephone-line streaming from the intern's computer to the university server where the supervisor was located was sometimes unreliable and at worst introduced a significant delay between voice and picture.

Gaining access to the appropriate equipment and materials was just the first step in ensuring that video supervision would provide the same opportunity to achieve equivalent results to face-to-face sessions. The supervisee also needed access to an on-site or close mentor to refer to between video sessions as a form of more immediate consultative support.

An extension of the use of video links for supervision sessions, especially applicable in training situations, was for the supervisor to have the supervisee demonstrate skills on video and send them to the supervisor prior to the online session. The supervisor could then preview and prepare questions ahead of the online session. Clips from the video could be shown during the session to provide a basis for discussion and could demonstrate where the supervisee showed particular strengths or elements of their work that required special attention.

A particularly useful advantage of the Web-based supervision, including the use of video links, was the ability to connect multiple supervisees in group supervision. Streaming multiple webcam images still remains problematic for the technology available today, although improvements are constantly appearing. However, their use in conjunction with other forms of distance contact, such as through email lists, scheduled chat sessions, weekly emailed reflective journals and bulletin board-style threaded discussions, worked well. In threaded discussions, supervisors pose questions that are developed to stimulate supervisees' thinking, while providing a window into the interns' understanding of the process.

Using this approach also encouraged the supervisor to be more aware of the day-to-day issues encountered by interns. In fact, the threaded discussion in the pilot seminar brought together eight interns, placed in six different districts (rural, suburban and urban, with one based 375 miles from the supervisor), for conversations that would not have occurred in an on-campus programme. Although distance supervision eliminates geographic barriers, it is vital to keep in mind that new barriers may be created. When servers go down, access to the supervisor is limited. In addition, supervisees must plan for the costs of Internet and telephone access.

Support for supervisees was another important aspect of offering such facilities. Examples of important information and technical support they required included:

- details of system requirements
- telephone numbers of supervisors and computer centre staff
- availability of a staffed help desk during day and evening
- training of supervisees on technological and software issues prior to the supervision
- Web software that is easy to load
- clear and uncluttered websites for the supervisor and supervisee.

Infrastructure and need are often something of a chicken and egg situation. There is no need to develop a large infrastructure if the extent of Web-based supervision does not warrant it. As the extent of video supervision increases, the infrastructure might lag behind need. Supervisors should expect growing pains and recognize that some of the burden will be on them to learn about the hardware, software and process necessary for supporting supervisees. At the same time, as more and more counselling supervisors begin using these methods, universities or other organizations housing these individuals should strive to provide additional technical assistance for both the supervisors using the Web-based setting and the supervisees.

TELEPHONE SUPERVISION

Research and even anecdotal literature on telephone supervision of clinical practice in the psychological therapies is remarkably rare. It is therefore necessary to rely largely on theoretical understanding and clinical experience as a foundation for discussion. There are many factors that can make face-to-face supervision impractical or otherwise not viable. Time, expense, travel and conflicting personalities where alternative supervisors are hard to find are but a few that come to mind. Especially in parts of the world where populations are widely scattered, it can be difficult to find a supervisor close at hand who suits the practitioner well. Other issues confront counsellors in rural and regional areas, such as potential threats to confidentiality arising from a possible overlapping of clients when their supervisor is the only other counsellor or therapist in the district. In these types of situation, telephone supervision is possibly a better alternative. Supervision should be appropriate to the circumstances of the practitioner and this may mean that finding an appropriate supervisor necessitates looking beyond a simple one-hour journey by car.

A supervisor who lives hundreds, if not thousands, of miles away from their supervisees is unlikely to face issues such as competition and confidentiality. The supervisee is not faced with the allied costs of time and travel. All this can help to offset the cost of long-distance telephone calls. Even practitioners who work in non-profit agencies need to be time and money smart, as they are still competing with other services for the public to use their facilities; otherwise funding is likely to be threatened.

In many parts of the world, it is commonly considered undesirable, or even unethical, for clinical supervision to be provided by those who also carry a line-management responsibility for the supervisee, for the sake of avoiding the obvious conflicts of interest that may arise. This has been found to be a major issue in, for example, military forces and police services. Telephone supervision conducted by an outsider is a very real alternative even where other sources of supervision separate from management is unavailable. Supervisees have been known to make comments such as: 'How freeing it is to talk to someone who is not involved in the machinations and politics of the workplace.' Again, for agencies and government organizations, telephone supervision can be a time-effective and employee-friendly alternative.

Many counsellors may be concerned about the restrictions of telephone supervision, such as not being able to read non-verbal communication and body language. This is a relevant point. However, in most cases a supervisee is paying good money for supervision and it is not in their own best interest to tell lies and not be open. Furthermore, the human voice can be difficult to mask and is often the first indicator of trouble. Non-verbals are often accompanied by sounds such as groans, deep exhaling or inhaling, breathing through the nose instead of the mouth or vice versa. These sounds can often inform the supervisor that something other than that which has been stated is wrong. An example is the supervisee who, after saying that everything was okay, exhaled deeply. Knowing that this was uncommon for this particular supervisee, the supervisor challenged him and found out that his brother had recently been killed in a car accident. It does not take long to identify what is normal in relation to a supervisee's expression over the telephone (see also Chapter 5).

ISSUES SPECIFIC TO TELEPHONE SUPERVISION

Telephone supervision is ideal for counsellors who practise in rural or isolated locations, bringing with it an issue that is unique to supervisors who offer their services to practitioners in other areas: the need for a very good knowledge of interstate and industry-relevant information. In Australia, for example, a supervisor who has supervisees outside his or her own state would need to be aware of issues such as the fact that in South Australia there is new legislation about mandatory reporting; in Queensland there is new legislation that requires all practitioners who see children to have a 'suitability card'; and in Victoria the new privacy legislation now requires practitioners to ensure that all clients fill out and sign consent forms. This is just a small sample of the knowledge required by telephone supervisors as a result of crossing geographical boundaries. This type of issue could be pertinent for supervisors in any nation that has federal and state laws or who work across national boundaries by any distance method.

Other issues can have a major impact for those working in areas of low population where maintaining a practice can be especially problematic. A telephone supervisor, or one working with supervisees at a distance by any means, needs to be more aware of specific issues that are geographically influenced, such as suicide. For instance, in Australia during the period from 1988 to 1998, persons living in capital cities had the lowest rates of suicide, persons living in urban areas had the second lowest rates of suicide, while rural areas had the highest suicide rates, ranging from 14.6 to 18.5 per 100,000 people (Australian Bureau of Statistics). Therefore, with fewer practitioners in rural areas and a higher suicide rate, it would be expected that rural services would be faced with this issue more often than their city counterparts. A city-based telephone supervisor would need to be aware of the higher suicide rate in rural areas and the relevant issues in relation to this subject, in order to ensure a satisfactory service to rural practitioners.

Another example is the supervisee whose client who was displaying distinct psychotic behaviour. The supervisee was concerned that the client needed help beyond their skills and that the client was also potentially dangerous. Being based in a city, the supervisor ignorantly suggested that he simply ring the local adult mental health team in his area and have his client assessed for hospitalization. The supervisee had already rung them and been informed that they were only interested if the client was actually hurting someone or was about to self-harm. It later transpired that there was only one team covering a distance of approximately 1,000 square kilometres (over 620 square miles). Due to this large coverage, they could not answer calls unless the situation was life-threatening. If they were answering a call that required many hours of travel, they were then unavailable for a significant period of time. If the caller could wait, they would book an assessment in the next couple of weeks. The supervisor's ignorance of a situation well-known to rural counsellors could have had detrimental consequences for the supervisee and his client, although in this instance it was possible to contain the situation until the client was assessed and ultimately hospitalized.

A respectful relationship needs to be formed between any supervisor and supervisee in regard to such things as confidentiality and bonding. Close professional relationships have been formed with supervisees using telephone supervision. One supervisee, in response to enquiries about the advantages and disadvantages as to the method, responded that he did not feel inhibited in being able to form a professional relationship because he felt safer on the telephone and had always felt as though he was under the microscope when he tried face-to-face supervision. Some practitioners feel more comfortable on the telephone than with face-to-face supervision and vice versa. In the experience of the authors, if a supervisee is comfortable, feels safe and is accorded respect, the mode does not detract from the ability to form a good working alliance.

CONCLUSION

Video-based and telephone supervision are now possible as real alternatives to more traditional styles. They are efficient, flexible, cost-effective and remove the stress of meeting appointment times that necessitate travel to the place of the appointment. Even before there is substantial evidence that particular models of supervision are more salient than others in a person-to-person format (Sexton, 1998), technology is causing supervisors, trainers and practitioners to rethink the direction that supervision is taking. However, research about relationship development via videoconferencing and telephone is needed to clarify the issues therein (Sampson et al., 1997). The question of how technology impacts on the quality of the counsellor–supervisor relationship must continue to be addressed in supervision research.

It is dangerous to fall victim to the idea that technology can somehow reduce the need for a knowledgeable supervisor who uses pedagogically sound principles of instruction and supervision. Reeves (1996) suggested that teachers abandon the perspective of 'technology as instructional media' and shift to viewing technology as something that they learn to use in instruction. This can also be applied to counselling supervision. It is not the medium that supervises the candidate; it is the supervisor. A supervisor's expertise, feedback and availability remain critical to effective supervision in any setting, whether it is through video links, the use of telephone supervision or any other means.

Web-based formats do not yet eliminate the need for mentoring by an experienced practitioner. Therefore, at the moment, practitioners who are not within a reasonable travelling distance of a willing mentor and who cannot attend in person for any supervision should the need arise may not be good candidates for a cyber-supervision setting. This is an example of the tension that can occur between the desire to improve access and the need to maintain high-quality supervision. It should also be noted that significant learning occurs as a result of interactions in hallways, between sessions and over coffee, an aspect that is not always available to those at remote sites. These types of interaction are often unscheduled but should not be regarded as incidental. In some settings they can be an important characteristic of supervision and can play a vital role in enculturation of a counsellor to become part of the professional community. Therefore, when this is absent at remote sites, the supervisee should be encouraged to join professional organizations and networks both online and offline.

As more training programmes and supervisors include video supervision and/or telephone supervision, the process of supporting practitioners in the profession will be altered. It will require supervisors to learn new ways to create caring communities that maximize the potential use of new technologies.

Bossert (1997) stated that, while education was relatively untouched by the revolution of transportation, it will be remarkably reformed by the current information and technological revolution that holds benefits and potential dilatory effects for students. The same can be said about counselling supervision. Bossert called on educators, as the authors of this chapter call on counsellor and therapist educators and supervisors, to be proactive in thinking about the potential impact of information technologies on this society. As Bossert (1997, p. 15) noted:

> The ancient Greeks had a saying that 'Fate leads the wise person by the hand, and drags the fool by the heels.' Everyone ends up at the same point in the end; it is just that some understand what is happening and take control of the situation, while others are dragged kicking and screaming into the future.

REFERENCES

Birchler, G.R. (1975) 'Live supervision and instant feedback in marriage and family therapy', *Journal of Marriage and Family Counseling*, **1**(4): 331–42.

Bloom, J.W. (1998) 'The ethical practice of Webcounseling', *British Journal of Guidance & Counselling*, **26**(1): 53–9.

Bossert, P. (1997) 'Horseless classrooms and virtual learning: Reshaping our environments', *Bulletin*, November, 3–15.

Bowman, J.T. (1980) 'Effect of supervisory evaluation on counsellor trainees' anxiety', *Psychological Reports*, **46**: 754.

Casey, J.A., Bloom, J.W. and Moan, E.R. (1994) Use of technology in counsellor supervision. In L.D. Borders (ed.) *Counseling Supervision*. Greensboro, University of North Carolina, ERIC Clearinghouse on Counseling and Student Services. (ERIC Document Reproduction Service No. Ed 372357.)

Hannon, K. (1996) 'Upset? Try cybertherapy', *U.S. News & World Report*, 13 May, p. 81.

Liddle, B.J. (1986) 'Resistance in supervision: A response to perceived threat', *Counsellor Education and Supervision*, **26**: 117–27.

Marino, T.W. (1996) 'Counselors in cyberspace debate whether client discussions are ethical', *Counseling Today*, January, p. 8.

Montalvo, B. (1973) 'Aspects of live supervision', *Family Process*, **12**(4): 343–59.

Neukrug, E.S. (1991) 'Computer-assisted live supervision in counsellor skills training', *Counsellor Education and Supervision*, **31**(2): 132–8.

Reeves, T. (1996) 'Relevant readings. Technology in teacher education: From electronic tutor to cognitive tool', *Action in Teacher Education*, **27**(4): 74–8.

Sabella, R.A. (1999) *SchoolCounselor.com: A Friendly and Practical Guide to the World Wide Web*. Educational Media, Minneapolis, MN.

Sampson, J.P. Jr, and Pyle, K.R. (1983) 'Ethical issues involved with the use of computer-assisted counseling, testing and guidance systems', *Personnel and Guidance Journal*, **61**: 283–7.

Sampson, J.P. Jr, Kolodinsky, R.W. and Greeno, P.B. (1997) 'Counseling on the information highway: Future possibilities and potential problems', *Journal of Counseling and Development*, **75**(3): 203–12.

Schnieders, H.L. (2000) From a bug in the ear to a byte in the eye: Implications for Internet delivered, live counsellor supervision. Monograph. http://cybercounsel.uncg.edu.

Sexton, T.L. (1998) 'Reconstructing counsellor education: Supervision, teaching, and clinical training revisited', *Counsellor Education and Supervision*, **38**(1): 2–5.

West, J.D., Bubenzer, D.L. and Zarski, J.J. (1989) 'Live supervision in family therapy: An interview with Barbara Okun and Fred Piercy', *Counsellor Education and Supervision*, **29**: 25–34.

Part III

Computerized therapy: stand-alone and practitioner-supported software

8 Computer programs for psychotherapy

KATE CAVANAGH, JASON S. ZACK,
DAVID A. SHAPIRO AND JESSE H. WRIGHT

INTRODUCTION

Computer systems are now widely used in psychotherapeutic practice to assist in assessment, diagnostic interviewing, consumer health education and practice management (see Bloom, 1992 and Burnett, 1989 for reviews). However, their most complex and controversial role is as psychotherapist, where the computer delivers a portion or the entirety of treatment. Until recently, computer-assisted therapy programs have rarely progressed beyond the prototype phase in which the software is tested in a small research study but never made available for widespread use. In this chapter we review the development of computerized psychotherapy from the pioneering work of the 1960s to current innovations that could significantly enhance the delivery of psychotherapy.

For four decades, researchers have attempted to identify the replicable ingredients of the psychotherapeutic encounter and translate these features into computer-delivered behavioural health interventions. Attempts to utilize computers in psychotherapy have been characterized by four waves to date, largely mirroring the zeitgeist in non-dynamic schools of psychotherapeutic thought:

1. client-centred or experiential (simulation of therapist–patient dialogue)
2. behavioural (training plus exposure or desensitization)
3. psychoeducational and cognitive interventions (programs that teach coping or problem-solving strategies, some of which employ cognitive restructuring)

4. cognitive-behavioural (more fully developed programs that contain a combination of methods typically employed in CBT and utilize multimedia or other contemporary technology).

The last three waves have been particularly suitable for computer adaptations because they are based on learning and behavioural theories, employ structured treatment methods and often contain specific interventions that can be translated to the computer format.

We have limited our review to those programs that are designed for the management of mental health problems, especially anxiety and depression. Excluded from this chapter is any discussion of computerized behavioural health programs targeting physical health problems or general health behaviours such as smoking or dieting, although some of the techniques and delivery mechanisms overlap between these and mental health programs. We have also eliminated from consideration any programs that have personal growth or enlightenment as their primary goals, as these activities are essentially different from counselling or psychotherapy per se and so warrant separate consideration. A discussion of the legal and ethical considerations of the implementation of computerized psychotherapies is also beyond the scope of this review (but see Chapter 9). Lastly, we do not discuss interactive voice response (IVR) systems developed for self-treatment of mental health problems (for a review of programs such as *COPE* and BT *STEPS*, see Marks et al., 1998).

THE FIRST WAVE: TICKERTAPE THERAPISTS

In 1966, computer scientists at Stanford University declared: 'We have written a computer program which can conduct psychotherapeutic dialogue' (Colby et al., 1966, p. 148). Their program was designed to 'communicate an intent to help, as the psychotherapist does, and to respond as he does by questioning, clarifying, rephrasing and occasionally interpreting' (p. 149). They conveyed a hope that this application might offer a widely available psychotherapeutic tool.

During the same year, Weizenbaum (1966), from the Massachusetts Institute of Technology, described a similar software program, 'ELIZA'. This system allowed people to discuss their problems using natural, ordinary words, phrases and sentence structure with their tickertape therapist. Weizenbaum considered his system a straw-man, a tool for exploring the possibilities and limitations of computer-simulated natural language using conversational rules based on non-directive, client-centred, Rogerian techniques featuring restatement and empathic reflection. He never intended ELIZA to function as a true psychotherapist (Weizenbaum, 1976). Primarily concerned with the ethical issues associated with computerized psychotherapy, he recommended against substituting a computer for a psychotherapist, whose role 'involves interpersonal respect, understanding, and love' (1976, p. 269).

Although a stimulus for great discussion, the use of natural language simulation in computerized psychotherapy has not been widely adopted in healthcare to date. Rather than offering empirical support for the use of such systems, computer scientists and psychotherapists alike detailed the limitations of these programs, emphasizing their shortfalls in generating meaningful therapeutic dialogue (for example Wagman, 1980; Wright and Wright, 1997).

After computerized psychotherapy's early foray into the simulation of natural language and psychotherapeutic dialogue, researchers argued that computerized psychotherapies were

> more attuned to brief, focused interventions of a cognitive or behavioural inclination, those that claim to be independent of non-specific aspects of therapy and that expect new relationships or behaviours to be experienced mainly outside the therapeutic setting. (Servan-Schreiber, 1986, p. 200)

Acknowledging their limitations, researchers focused on the translation of specific cognitive and behavioural techniques, methods based on learning theory, or problem-solving interventions to be delivered by computer systems.

THE SECOND WAVE: SIMPLE BEHAVIOURAL TECHNIQUES

Behavioural psychotherapies have enjoyed considerable success in the treatment of anxiety and are recognized as a treatment of choice for specific fears by health professionals (Roth and Fonagy, 1996; Department of Health, 2001). In combating specific fears, graded exposure with relaxation training is a highly effective and rapid intervention. The technique, ubiquitous in behavioural treatment of anxiety disorders, involves identifying and encountering a hierarchy of feared stimuli until no further anxiety is experienced in their presence. The structured and systematic nature of this psychotherapeutic technique lends itself particularly well to computerization, a key principle that has been seized upon by the authors of therapeutic computer programs. This section describes a number of programs delivering this technique and evaluates their effectiveness in the management of agoraphobia and other anxiety disorders.

Automated graded exposure

The earliest computerized desensitization programs were automated prototypes. The first automated desensitization procedure to be evaluated in a controlled trial was a system called DAD, developed by Lang et al. (1970). Following four sessions of therapist-guided relaxation training, the DAD system presented a 20-item snake-fear hierarchy, where exposure and relaxation episodes were controlled by the snake phobic participants using switches attached to the arms of their chair. Outcome measures indicated marked improvement in participants using the DAD system. Results were comparable to those who received therapist-guided desensitization. Indeed, on some measures DAD outcomes excelled over live therapy, leading the authors to conclude that live therapist involvement in desensitization may not only be an unnecessary luxury, but a hindrance.

Computerized graded exposure: standardized hierarchies

The first truly computerized graded exposure program was developed by Biglan et al. (1979) for the management of test-anxiety. Their program included audio-taped relaxation training and a computer-controlled desensitization system. Prior to using the program, students were trained to relax. In the computer-controlled program, a standard hierarchy of 20 textual descriptions of items related to test-anxiety was presented on a video display terminal. Test-anxious students were instructed to imagine the item while relaxing for a 30-second period. If during that time, the students felt discomfort, they pressed a key to indicate this. The computer then presented the instruction 'RELAX' and waited 30 seconds before re-presenting the item. If no discomfort was reported, then after 30 seconds of further relaxation the next item in the hierarchy was presented. Nine college students reporting marked test-anxiety completed the program in an average of four sessions lasting about 30 minutes each. Their scores on the test-anxiety scale demonstrated significant reduction from pre- to post-treatment as well as increased comfort in various aspects of the academic examination process.

Computerized graded exposure: individualized hierarchies

Expanding on this early work, two groups of researchers (Ghosh et al., 1984; Chandler et al., 1986; Ghosh and Marks, 1987) developed generic computer programs for systematic desensitization that permitted both individual behavioural treatment goals and individualized sensitization hierarchies. Both programs also expanded the locus of the therapeutic setting, incorporating both computer-based imaginal exposure techniques with computer-guided *in vivo* (live) exposure homework practice, supported by printed information summaries and diary sheets.

Chandler et al. (1986) presented a case report of a 35-year-old male diagnosed with agoraphobia and obsessive-compulsive ruminations. He reported being unable to leave his apartment alone unless he was going directly to his parents' house or other safe places. The computer program presented graded exposure in four stages. The first stage, delivered in session one, taught the patient the learning theory perspective on the aetiology of phobias, offered hope, presented the structure of the program and emphasized the patient's responsibility for, and importance of, completing the homework. This stage also included relaxation training delivered by audio tape. The patient was then instructed to practise the relaxation techniques daily before the next computer session. Practice was supported by printed instructions. The second stage involved the construction of a personalized phobic hierarchy. Third, the presentation of items from the hierarchy and imaginal exposure commenced, following a deep relaxation exercise. Four items from the hierarchy were presented in each session. The fourth stage, *in vivo* exposure, formed between-session homework. The extent to which imaginal exposure had been endured determined the program's recommendations for facing the hierarchy items in real life. During 13 sessions with the computer program, the patient improved dramatically, achieving therapeutic targets and life goals which were retained, indeed improved, at eight months' follow-up. From not being able to leave his apartment, he was able to set up in business as a tradesman working on external jobs, and able to go where he wanted to by himself. Equally dramatic improvements were seen in a case series of five patients suffering from a range of phobias (Chandler et al., 1988).

Ghosh and Marks (1987; see also Ghosh et al., 1984) reported on a randomized controlled study of 40 agoraphobics which investigated the effectiveness of their graded exposure therapy, compared to computerized instruction with psychiatrist-led therapy and a self-help book. Following a 90-minute assessment with a psychiatrist, patients either met weekly with the psychiatrist for exposure instructions, received the book *Living with Fear* (Marks, 1978), or planned their exposure treatment by interacting with the computer program. All three groups of agoraphobics improved substantially up to six months' follow-up, with no significant differences between them. As a group, the patients changed from being 'habitual avoiders with regular phobic panic pre-treatment, to becoming non-avoiders with no phobic panic and only residual slight anxiety at follow-up' (Ghosh and Marks, 1987, p. 9). On the basis of these research findings, Bloom (1992) argued that graded exposure 'does not appear to require interpersonal interaction with a therapist in order to be successful' (p. 183).

Marks et al.'s program (1998) has recently been updated technologically to include screen voice-overs and permit Web delivery. Research evidence also suggests that the updated program *FearFighter* is efficacious and cost-effective in the self-treatment of specific fears and agoraphobia (Shaw and Marks, 1996; Kenwright et al., 2001).

Computerized graded exposure: virtual reality

The use of virtual therapeutic environments has enjoyed some success in the treatment of spider phobia (Carlin et al., 1997), claustrophobia (Botella et al., 1999; Wiederhold and Wiederhold, 2000), acrophobia (Rothbaum et al., 1995a and b), social phobia (North et al., 1998; Petraub et al., 2001a; Wiederhold and Wiederhold, 2000), agoraphobia (Wiederhold and Wiederhold, 2000), fear of flying (Muehlberger et al., 2001; Rothbaum et al., 2001) and PTSD (Rothbaum et al., 1999, 2001).

Virtual reality offers a new interactive therapeutic paradigm in which users are no longer simply onlookers of images on a computer screen but become active participants within a computer-generated, three-dimensional virtual world. In virtual reality, real-time computer graphics, body-tracking devices, visual displays and other sensory input and feedback devices are integrated to give the user a sense of immersion or presence in the virtual environment (Kalawsky, 1993). Participants usually wear a head-mounted display presenting a virtual environment display that moves, naturally, in response to head and body motion. Sensor gloves and other features can be used to facilitate and enhance environmental interactions appropriate to the exposure encounter.

In the case of acrophobia, Rothbaum's group has developed virtual reality hardware and software to recreate the experience of a variety of elevation scenes. For example, footbridges above water, outdoor balconies and a glass elevator simulating the one at the Atlanta Marriot

Marquis convention hotel, rising 49 floors, up to 147 metres at the top. Desensitization conducted during immersion in these environments resulted in substantial incremental improvement compared to a waiting list control group. Following treatment, patients reported significantly reduced fear and avoidance of heights and improved attitudes towards heights. Similar benefits for flight-fearful individuals were observed in a controlled trial in which 49 patients with fear of flying were randomly assigned to virtual reality exposure and desensitization to a flight-fear hierarchy, standard therapist-guided exposure or a waiting list control group. Rothbaum et al. (2001) found that both virtual and standard exposure were superior to the waiting list control group in reducing fear of flying and achieving behavioural flight goals and that these improvements were retained to 12 months' follow-up. In the year after the study, 92 per cent of the virtual exposure group and 91 per cent of the standard, *in vivo*, exposure group took a flight.

The computer generation of virtual reality environments may be more acceptable to anxious clients than *in vivo* exposure (see Garcia-Palacios et al., 2001) and also permits much more control over the exposure episodes than is usually possible (North et al., 2002). It may offer opportunities for reducing the costs of therapist-guided exposure to stimuli which typically require leaving the therapist's office (heights, driving, shopping centres) and exposure episodes which can be costly and complex in themselves (social performance situations, flights, snakes, tarantulas). However, virtual reality equipment can be expensive. Currently these technologies are not widely available for clinical use.

In the case of social and performance anxiety, future virtual therapeutic environments may be augmented with avatars (virtual people) to create the illusion of an interactive audience. Petraub et al. (2001b) recreated a typically feared performance anxiety situation by creating a virtual audience of eight male avatars sitting in a semi-circle. In order to foster the illusion of life, the avatars exhibited small twitching movements, blinking and shifting about in their chairs. Audience behaviour was presented in two conditions by manipulating facial animations, direction of gaze, posture and physical animation displaying classic non-verbal communications of friendly receptivity and hostility. They found that participants, particularly in a high immersion group who interacted with the audience via a head-mounted display (rather than simply a desktop screen), responded to virtual audiences much as they would respond to real audiences. The authors suggest that this indicates that virtual reality scenarios with computer-generated characters could be of use in treating and investigating a range of social performance situations.

Moreover, virtual environments can be used to construct and explore feared or desired spaces that cannot be encountered in real life. Using these techniques, it is possible to transcend the limits of the real world, permitting the shared construction and exploration of memorial, imaginal or anticipated events (Riva et al., 2001). For example, virtual reality exposure and desensitization has been effectively used to reduce symptom severity associated with specific traumatic experience in Vietnam veterans using immersion into virtual environments such as flying in a Huey helicopter over virtual Vietnam and walking in a jungle clearing (Rothbaum et al., 2001).

THE THIRD WAVE: COMPUTER PROGRAMS BASED ON CBT, PSYCHOEDUCATIONAL, OR PROBLEM-SOLVING APPROACHES

The third wave of computerized therapy systems was characterized by the translation of cognitive-behaviour therapy, psychoeducational and problem-solving interventions to the computer format. In these programs, the computer therapist typically adopts a role of instructor and advisor-functions that are more easily embodied in a computerized therapy paradigm than those of a psychodynamically oriented therapist (Wagman, 1980). The first computer programs in this wave were based on psychoeducational or problem-solving interventions. Later programs incorporated at least some of the basic methods of CBT. However, none of the programs in this group presented a fully developed method of using interactive CBT self-help exercises.

Dilemma Counselling System

Whereas Weizenbaum (1966) and Colby et al. (1966) had attempted to mimic the most complex and sophisticated forms of psychotherapy, Wagman championed exploration of a more simple approach. With his PLATO Dilemma Counselling System (DCS), Wagman (1980) targeted a single technique in psychotherapeutic interaction: dilemma counselling. The systematic dilemma counselling paradigm assumes that discomfort arises from a person's need to choose between what are thought of as two undesirable alternatives, or avoidance-avoidance problems. For example, in the case of career choice for university students, Wagman (1980) showcased the dilemma that choosing finance is a lucrative but intellectually unstimulating option, and choosing to be a scientist is intellectually stimulating but relatively unlucrative. The PLATO DCS presents an overview of the rationale for the dilemma counselling method and then offers five stages in resolving each dilemma:

1. it aids the user in formulating the problem as a psychological dilemma
2. it helps the user to develop an extrication route for each dilemma
3. it helps the user to identify a line of enquiry which will help negotiate each extrication route
4. it helps the user to generate solutions for each line of enquiry
5. it guides the patient to rate and evaluate the various proposed solutions.

Early evaluation of the DCS method indicated that it could be valuable for people with a troublesome but clearly articulated dilemma. In a randomized controlled trial, students using DCS reported greater problem improvement one week and one month after using the system than control subjects. Ninety per cent of those using the system found it to be at least slightly helpful. Moreover, the majority of participants felt that the system was not too impersonal and it was stimulating and interesting. Many participants felt more at ease using the computerized system than seeing a human counsellor (Wagman, 1980; Wagman and Kerber 1984).

Therapeutic Learning Program

Colby et al. (1989) have described a method of short-term computer-assisted psychotherapy for anxiety problems that functions adjunctively to group therapy sessions. The *Therapeutic Learning Program* (TLP) requires ten hours of group work in which six to ten people each work independently on computerized psychotherapy systems and then discuss their individualized printout with the group. According to Colby et al. (p. 105) the sessions involve eight stages:

1. Identify the demand inherent to the patient's interpersonal problem situation that is not being addressed effectively
2. Identify new proactive behaviour (action steps) that might effectively address the dissatisfied state

3. Clarify the suitability of the new proactive behaviour and identify the inhibited function that results from the patient's prediction of feared adverse consequences
4. Identify the incorrect beliefs (thinking errors) that link catastrophic predictions to an action intention
5. Help the patient to understand the historical origin of thinking errors in childhood adaptations and sort out present realities from past realities
6. Help the patient understand that childhood thinking is no longer appropriate for adult decision-making in interpersonal problem situations
7. When the predicted adverse consequences are accepted as incorrect, the patient's fear of transgressing rigid command rules diminishes and the patient is more likely to carry out the required proactive behaviour
8. The recovery of an inhibited function becomes part of the patient's self-concept as a functioning adult.

Positive early reports of subjective experiences in using this technique (Talley, 1987) led to a later study including almost 300 participants. Over 95 per cent reported improvement in their ability to handle the problem situation that brought them to the therapy, 78 per cent reported a drop in general distress and 78 per cent reported a high level of satisfaction with the procedure (Dolezal-Wood et al., 1996).

Another approach to offering computerized advice and psycho-education for problems of anxiety was reported in Parkin et al. (1995). The prototype program 'WORRYTEL' was designed to guide users to develop their own self-treatment program, permitting them to view case vignettes, identify personal symptoms and print out useful screens from the program. A small pilot study of the program suggested that anxious patients found it easy to use, acceptable and empathetic, but wanting more specific guidance in how to overcome their current problems.

Cognitive-behavioural therapy for panic

A small randomized controlled trial by Newman et al. (1997) demonstrated that a palmtop therapeutic aid could increase the cost-effectiveness of cognitive therapy for panic. The traditional therapy group received 12 sessions of individual CBT, whilst the palmtop group received just four individual therapy sessions and completed the remainder of the course guided by their palmtop panic program. Both groups benefited from the therapy. However, subjects who received the standard treatment had significantly greater improvement at the end of active treatment. Therapeutic gains converged during the follow-up period with no significant differences found between computer-assisted CBT and standard CBT six months after completion of the treatment. The authors estimated that the palmtop treatment saved US$540 (approximately £330 at the time of writing) per panic treatment which, in the face of equivalent clinical outcomes, represented a considerable potential for healthcare cost saving.

Cognitive therapy for depression

Selmi (1982) developed an early computerized psychotherapy program based on the work of Beck (1976; Beck et al., 1979). Selmi et al. (1990) reported evidence for the efficacy of this form of computerized cognitive therapy for depression. Thirty-six patients meeting clinical criteria for depression were randomly assigned to therapist-led psychotherapy, computerized psychotherapy or waiting list control. Both active treatments were found to be superior to the waiting list control group. The Selmi et al. program relied extensively on text to communicate with the user and did not incorporate multimedia elements. Although it is no longer available for clinical use, this software demonstrated the feasibility of using a therapeutic computer program to deliver some of the basic elements of cognitive therapy.

Overcoming Depression

Colby's program *Overcoming Depression* incorporated two quite different modules. The first module included a natural language component that attempted to simulate a non-directive therapeutic interview. The second module included psychoeducational material based in part on CBT. Responses were entered with a keyboard or mouse; and multimedia was not used, thus the program relied heavily on the user's ability to read and understand written text. A study conducted at the University of Iowa found that *Overcoming Depression* did not significantly help depressed inpatients and was significantly less useful than cognitive therapy delivered by a human therapist (Bowers et al., 1993). In this study, subjects had negative responses to the natural language module (Stuart and LaRue, 1996). The investigators noted that the computerized interview was often off target or seemed to misunderstand the communications of the user. Thus, patients became frustrated with the software. In contrast to the research at the University of Iowa, positive subjective responses were reported in an uncontrolled study by Colby (1995). Randomized, controlled trials with depressed outpatients have not been completed.

THE FOURTH WAVE: MULTIMEDIA INTERACTIVE COGNITIVE-BEHAVIOURAL THERAPY FOR DEPRESSION AND ANXIETY

Cognitive-behavioural therapies have become increasingly important treatments for depression and anxiety because of their demonstrated efficacy (Roth and Fonagy, 1996; Department of Health, 2001). A primary strategy in this approach is to analyse the interaction of thinking styles and behaviour patterns so that clients can identify antecedents to distressing emotional states and modify them. Recently, two groups of developers have been working on fully developed, computerized CBT programs that use multimedia and other contemporary computer tools to heighten the learning experience (Wright et al., 1995, 2002a; Proudfoot et al., 2003a and b).

The newest computerized therapy systems are designed to embody both the specific active techniques of this therapeutic approach and the non-specific features of the therapeutic relationship known to influence clinical outcomes (for example alliance or engagement, empathy, motivation and trust). The multimedia format offers a stimulating and engaging interface, integrating video, graphics and animations, voice-over and many interactive episodes, including multiple choice responding, distress/success ratings, on-screen problem-solving and diary completion and so on.

Two empirically supported interactive multimedia systems are currently available: *Good Days Ahead: The Multimedia Program for Cognitive Therapy* (USA, Wright et al., 2002a and b) and *Beating the Blues* (UK, Proudfoot et al., 2003a and b). The former is designed primarily to be used under the supervision of a clinician as part of an integrated package of computer-assisted CBT. However, an edition of the software is also available that can be used independently as an electronic self-help method. *Beating the Blues* has been designed mainly as a stand-alone tool.

Good Days Ahead: The Multimedia Program for Cognitive Therapy

The first multimedia program for CBT was developed by Wright and co-workers (1995) and released in a prototype laser disc version (titled *Cognitive Therapy: A Multimedia Learning Program*) in 1995. An upgraded, DVD-ROM version of this software (*Good Days Ahead*) was subsequently produced and is currently available for clinical use (Wright et al., 2002a). The content for this software, authored by highly experienced cognitive therapists, covers the core self-help methods of CBT. Program modules include topics such as 'Basic Principles' (for example psychoeducation on the basic CBT model, introduction to methods of changing thoughts and behaviour); 'Changing Automatic Thoughts' (for example identifying automatic thoughts, spotting cognitive errors, examining the evidence, developing rational alternatives and thought recording); 'Taking Action' (for example activity scheduling, pleasant events scheduling, graded task assignments and hierarchies); and 'Changing Schemas' (for example identifying and modifying basic beliefs and attitudes).

Video and audio are used liberally throughout this program to illustrate how the characters (portrayed by professional actors) use CBT methods to overcome depression and anxiety. A variety of interactive self-help exercises, graphics, multiple choice questions, checklists and mood ratings are used to engage the user and encourage use of CBT methods in real-life situations. Homework is assigned after each session on the computer, and feedback is given to reinforce learning. Although a keyboard is used to enter some responses on self-help exercises, much of the program can be performed by making selections with a mouse. Program content is written at the ninth grade reading level (around age 14–15) and is designed to be suitable for persons with no previous computer experience.

In an uncontrolled, preliminary trial, 96 inpatients and outpatients, most diagnosed with major depression, were permitted to use the program at their own pace in conjunction with any available treatment as usual (typically a mixture of pharmacotherapy and psychotherapy). Users indicated a high rate of acceptance of this form of computer-assisted therapy, with mean satisfaction scores of about 4.5 on a 5-

point scale (5 = highest rating). The entire program was completed by 78.1 per cent of subjects and 93.4 per cent reached at least the midpoint. Mean scores on a measure of cognitive therapy knowledge were significantly improved (Wright et al., 2002b). This open trial was not designed to test the efficacy of computer-assisted CBT.

A subsequent randomized, controlled study has supported the clinical efficacy of this program (Wright et al., 2001). Drug-free subjects (n = 45) with major depression were randomly assigned to computer-assisted cognitive therapy (CCT), standard cognitive therapy (CT), or a waiting list control group (WL). Treatment consisted of up to nine sessions over eight weeks. CCT involved abbreviated visits with a clinician (25 minutes instead of 50 minutes in CT) plus use of the computer program (about 25-minute sessions). Results showed that CCT was equal to CT and both were superior to the WL. The response rate (50 per cent drop in BDI scores) was identical in CCT and CT (70 per cent), and was only 7 per cent in WL subjects. Treatment gains were well maintained at the three and six months' post-treatment evaluations. Depression scores were unchanged from those recorded at the end of treatment for both active therapies.

Beating the Blues

Beating the Blues offers cognitive-behavioural therapy for anxiety and depression in the form of a stand-alone, computer-controlled, interactive multimedia package which can be delivered on a personal computer located in the primary or psychotherapy care practice or community resource centre. Clinical supervision and responsibility continue to rest with the primary care physician or other appropriately qualified personnel (nurse or clinical psychologist), to whom reports (including warnings of suicide or other risk) are automatically delivered by the computer program.

Beating the Blues is an eight-session program (each session lasting about one hour), plus homework assignments between sessions. The program is readily usable by patients with no previous computer experience. Like other versions of cognitive-behavioural therapy, *Beating the Blues* can be given alone or in combination with pharmacotherapy. *Beating the Blues* utilizes a range of multimedia capabilities. It features a series of filmed case studies of fictional patients who are used to model both the symptoms of anxiety and depression and their treatment by cognitive-behavioural therapy, as well as animations, voice-over and interactive modules.

In a recent study, researchers used *Beating the Blues* with a sample of 170 patients suffering from anxiety, depression, or mixed anxiety and depression (Proudfoot et al., 2003a) The randomized controlled trial demonstrated the clinical benefits of *Beating the Blues*. Patients were first of all prescribed treatment as usual (TAU) by their primary care doctor (physician) – whatever the doctor regarded as appropriate for that patient. They were then randomized separately in two categories, drug or no drug, depending on the physician's prescribed treatment, to TAU alone or TAU plus *Beating the Blues*. All patients completed questionnaires (including the Beck Anxiety and Depression Inventories) at entry to the trial, two months later (when patients randomised to *Beating the Blues* had completed the program) and at one, three and six months' follow-up.

At entry, levels of anxiety and depression were in the high moderate to severe range and comparable to those observed in other studies of general practice. As expected, patients allocated by their doctor to drug treatment were more severely ill. The prescribed pharmacotherapy was effective on all measures. Also as expected, patients who, at entry to the trial, had been ill for six months or more were more anxious and depressed than those with shorter periods of pre-existing illness.

The most important finding was that computer therapy reduced anxiety and depression significantly and substantially: average scores at the end of treatment in patients allocated to *Beating the Blues* were barely above the non-clinical range. These positive effects remained stable through follow-up to the final six months' measurement. Moreover, they did not interact with drug treatment: the beneficial effects of drug and *Beating the Blues* were additive. Nor did the effects of *Beating the Blues* interact with either the severity or duration of pre-existing illness. Notably, patients more than six months' ill at the start of treatment responded as well to computer therapy as did those with shorter periods of illness.

In sum, research findings indicate that computer-delivered cognitive-behavioural therapy can be a valuable tool for the treatment of anxiety and depression across a wide range of patients seen with these conditions in general practice (Proudfoot et al., 2003a) as well as for inpatients and outpatients within the context of psychiatric practices (Wright et al., 2001, 2002b).

Below we include a case study to illustrate the experience of a client using a current computerized psychotherapy package.

case study

Janet was a 35-year-old woman who sought treatment for depression after the break-up of her marriage. She and her ex-husband had been co-owners and operators of a small, but successful business. About six months before entering treatment, she discovered that her husband was having an affair. When she confronted him, Janet was bombarded by a litany of criticisms and complaints. Her husband, Donald, blamed the entire problem on Janet. After accusing her of paying so much attention to the business that she was ignoring him, Donald delivered a stinging rebuke. He told her that he had never really loved her and their entire marriage had been a sham. Within a few days, he filed for divorce.

By the time Janet came for an initial consultation, her self-esteem was at an all-time low. She was severely depressed, had great difficulty in going to work and wasn't participating in any of her usual social and recreational activities. Janet had been exercising regularly before becoming depressed, but hadn't done any exercise for the past three months. Sleep, appetite and energy levels also were disturbed. Her thinking had turned in a strongly negative direction. Typical automatic thoughts were: 'No one would ever want me…I'm a loser…I failed at my marriage…I was the one who messed up…I didn't try hard enough…I'll be miserable the rest of my life…I don't do anything right.' Although Janet had become hopeless and had thoughts of wishing to die, she reported that she would never kill herself because of her two children.

After an initial evaluation lasting approximately one hour, Janet started a course of computer-assisted cognitive therapy (CCT) using software developed by Wright and co-workers. She used the laser disc version of software (*Cognitive Therapy: A Multimedia Learning Program*) that has recently been released in an upgraded, DVD-ROM edition (*Good Days Ahead*). Although antidepressant therapy also was recommended, Janet decided against using psychotropic medication. Five brief sessions (20 minutes each) were held with a clinician who reviewed homework from the computer program, reinforced the use of cognitive therapy skills, answered questions about the computer-assisted treatment and helped to customize computer-based learning to meet the needs of the patient.

Janet worked at her own pace in using the therapeutic software in sessions typically lasting 25–50 minutes. She also did homework in a companion workbook that she brought to therapy sessions to discuss with the clinician. The total amount of therapy time, including the initial evaluation, was about 150 minutes with the clinician and about 240 minutes with the computer program (a total of 6.5 hours of clinician plus computer time).

In initial sessions of CCT, Janet learned the basic cognitive-behavioural model, including the relationship between dysfunctional automatic thoughts, dysphoric emotions and maladaptive behaviour. The therapeutic software demonstrated the basic cognitive model by showing videos from the life of the main character, a woman suffering from depression and anxiety. In early segments Janet viewed the main character in scenes where negative automatic thoughts led to depressed and anxious moods and defeated, helpless behaviour. After seeing these vivid demonstrations, Janet answered questions to reinforce learning, received computerized feedback and began to identify her own negative automatic thoughts. Her response to this software was one often seen: there was a rather sudden, almost revelatory experience in which she identified with the main character and realized that her own dysfunctional thinking was a large part of her problem. A great deal of emotional relief followed shortly thereafter.

Janet soon recognized that her own thinking was riddled with negative automatic thoughts and cognitive errors. In the third segment of the software (Changing Automatic Thoughts) she learned how to spot cognitive errors, do thought recording, examine the evidence for automatic thoughts and develop rational alternatives. After using this segment of the computer program, Janet brought several five-column thought records to her next visit with the clinician. These thought records contained a number of examples of cognitive restructuring that were stimulated by the computerized component of the therapy. For example, Janet noted that thoughts such as 'No one will ever want me' and 'I was the one who messed up' were full of cognitive errors such as ignoring the evidence, magnifying and personalizing.

Her clinician was able to capitalize on the positive learning experiences from the computer program in order to help Janet see how she had taken almost complete blame for the divorce, was forgetting her strengths and was forecasting an unnecessarily bleak future. He suggested that in addition to the homework from the computer program, Janet might try to rekindle relationships with some of her old friends so that she could begin to refocus her life away from the divorce-related grief.

Subsequent modules from the therapeutic software were used to help Janet break out of her pattern of helpless, isolated behaviour and recognize underlying schemas that predisposed her to react so strongly to the divorce. In the Taking Action segment she learned how to use activity scheduling, mastery and pleasure ratings and graded task assignments to reverse depressive behavioural patterns. The Changing Schemas module helped her to uncover some of her underlying maladaptive and adaptive attitudes and learn methods of changing these beliefs. One of the most damaging negative schemas was 'Without a man (that is, my husband), I'm nothing.' After recognizing this core belief, Janet was able to examine the evidence and find that she actually had many strengths, interests and potentials that were fully independent of her relationship with her husband or any other man. For example, she was the principal developer of their business and had been the primary reason for its success.

Janet's depression resolved rapidly with computer-assisted cognitive therapy. Within two weeks, her depressive symptoms had fallen by at least 50 per cent. She was much more hopeful and had no more thoughts of wishing she were dead. Her helpless and isolated behaviour also began to change early in the treatment process. After one week, she was back at work full time and was starting to exercise again. By the end of six weeks of treatment, she had no residual symptoms of depression. Self-esteem was much improved, and she had made significant strides in re-establishing her social life. Janet was seen once for a follow-up visit two months after completing computer-assisted therapy. She was doing well, and reported that she was still using cognitive therapy skills in daily life.

WHAT WORKS FOR WHOM?

The landscape of mental health service provision varies from country to country, state to state, and indeed, town to town. One shared feature is that national health services, managed care organisations, employee assistance providers, individual clinicians, community mental health centres and other mental healthcare providers are all seeking opportunities to increase access to more acceptable, more beneficial and more cost-effective early interventions for mental health problems.

A review of the research literature regarding computerized psychotherapies indicates that where proven techniques are translated for computer delivery clinical outcomes are comparable to traditional face-to-face services and may be indicated for both anxiety and depression. The best supported programs meet recognized standards of proof for psychotherapeutic interventions (Department of Health, 2001), as well as independent acknowledgement for the benefits of disseminating these programs within healthcare systems (Kaltenthaler et al., 2002).

Little is known about which patient characteristics predict most engagement or benefit from computerized psychotherapy programs. Research is currently underway to identify such features.

THE FUTURE

This chapter offers a snapshot of a very young set of technologies: a brief review of a few dozen papers which largely constitute the published empirical literature regarding computerized psychotherapies. Today, at a confluence of technological capability and societal acceptance of computers as integral to daily life, true computerized psychotherapies have just begun. The published literature reviewed here is not wholly representative of the state of the field, as empirical psychotherapy research often follows in the footsteps of clinical pioneers. However, the majority of published research suggests that computerized psychotherapies:

- offer an acceptable format for care
- increase access to effective psychotherapies for people suffering from anxiety and depression
- reduce symptoms and problem severity
- improve functioning and wellbeing
- reduce the costs of delivering proven techniques in mental healthcare.

REFERENCES

Beck, A.T. (1976) *Cognitive Therapy and the Emotional Disorders*. Penguin, London.

Beck, A.T., Rush, A.J., Shaw, B.F. and Emery G. (1979) *Cognitive Therapy for Depression*. Guilford, London.

Biglan, A., Villwock, C. and Wick, S. (1979) 'Computer controlled program for treatment of test anxiety', *Journal of Behaviour Therapy & Experimental Psychiatry*, **10**: 47–9.

Bloom, B.L. (1992) 'Computer-assisted psychological intervention: Review and commentary', *Clinical Psychology Review*, **12**: 169–97.

Botella, C., Villa, H., Banos, R., Perpina, C. and Garcia-Palacios, A. (1999) 'The treatment of claustrophobia with virtual reality: changes in other phobic behaviors not specifically treated', *CyberPsychology and Behavior*, **2**: 135–41.

Bowers, W., Stuart, S., MacFarlane, R. and Gorman, L. (1993) 'Use of computer-administered cognitive-behaviour therapy with depressed inpatients', *Depression*, **1**: 294–9.

Burnett, K.F. (1989) 'Computers for assessment and intervention in psychiatry and psychology', *Current Opinion in Psychiatry*, **2**: 780–6.

Carlin, A.S., Hoffman, H.G. and Weghorst, S. (1997) 'Virtual reality and tactile augmentation in the treatment of spider phobia: A case report', *Behaviour Research and Therapy*, **35**: 153–8.

Chandler, G.M., Burck, H.D. and Sampson, J. (1986) 'A generic computer program for systematic desensitisation: description, construction and case study', *Journal of Behaviour Therapy & Experimental Psychiatry*, **17**: 171–4.

Chandler, G.M., Burck, H.D., Sampson, J. and Wray, R. (1988) 'Effectiveness of a generic computer program for systematic desensitisation', *Computers in Human Behaviour*, **4**: 339–46.

Colby, K.M. (1995) 'A computer program using cognitive therapy to treat depressed patients', *Psychiatric Services*, **46**: 1223–5.

Colby, K.M., Gould, R.L. and Aronson, G. (1989) 'Some pros and cons of computer-assisted psychotherapy' *Journal of Nervous and Mental Disease*, **177**: 105–8.

Colby, K.M., Watt, J.B. and Gilbert, J.P. (1966) 'A computer method of psychotherapy: Preliminary communication', *Journal of Nervous and Mental Disease*, **142**: 148–52.

Department of Health (2001) *Treatment Choice in Psychological Therapies and Counselling: Evidence Based Clinical Practice Guidelines*. Department of Health, London.

Dolezal-Wood, S., Belar, C.D. and Snibbe, J. (1996) 'A comparison of computer-assisted psychotherapy and cognitive behavioral therapy in groups', *Journal of Clinical Psychology in Medical Settings*, **5**: 103–15.

Garcia-Palacios, A., Hoffman, H.G., Kwong See, S. et al. (2001) 'Redefining therapeutic success with virtual reality exposure therapy', *CyberPsychology and Behavior*, **4**(3): 341–8.

Ghosh, A. and Marks, I.M. (1987) 'Self-treatment of agoraphobia by exposure', *Behaviour Therapy*, **18**: 3–18.

Ghosh, A., Marks, I.M. and Carr, A.C. (1984) 'Controlled study of self-exposure treatment for phobics: preliminary communication', *Journal of the Royal Society of Medicine*, **77**: 483–7.

Kalawsky, R.S. (1993) 'VRUSE – a computerised diagnostic tool: For usability evaluation of virtual/synthetic environment systems', *Applied Ergonomics*, **30**: 11–25.

Kaltenthaler, E., Shackley, P., Stevens, K., Beverley, C., Parry, G. and Chilcott, J. (2002) 'Computerised cognitive behavioural therapy for depression and anxiety'. Unpublished manuscript.

Kenwright, M., Liness, S. and Marks, I.M. (2001) 'Reducing demands on clinicians time by offering computer-aided self help for phobia/panic: feasibility study', *British Journal of Psychiatry*, **179**: 456–9.

Lang, P.J., Melamed, B.G. and Hart, J.A. (1970) 'Psychophysiological analysis of fear modification using an automated desensitisation procedure', *Journal of Abnormal Psychology*, **76**: 220–34.

Marks, I.M. (1978) *Living with Fear: Understanding and Coping with Anxiety*. McGraw-Hill, London.

Marks, I.M., Shaw, S. and Parkin, R. (1998) 'Computer-aided treatments of mental health problems', *Clinical Psychology: Science and Practice*, **5**: 151–70.

Muehlberger, A., Herrmann, M.J., Wiedemann, G., Ellgring, H. and Pauli. P. (2001) 'Repeated exposure of flight phobics to flights in virtual reality', *Behaviour Research and Therapy*, **39**: 1033–50.

Newman, M.G., Consoli, A.J. and Taylor, C.B. (1997) 'Palmtop computer program for treatment of GAD: development of a cost-effective psychotherapy aid', *Behaviour Modification*, **32**: 11–18.

North, M.M., North, S.M. and Coble, J.R. (1998) 'Virtual reality therapy: An effective treatment for phobias', In Riva, G. and Wiederhold, B.K. (eds) *Virtual Environments in Clinical Psychology and Neuroscience: Methods and Techniques in Advanced Patient–Therapist Interaction. Studies in Health Technology and Informatics*, pp. 112–19. IOS Press, Amsterdam, Netherlands Antilles.

North, M.M., North, S.M. and Coble, J.R. (2002) 'Virtual reality therapy: An effective treatment for psychological disorders' In Stanney, K.M. (ed.) *Handbook of Virtual Environments: Design, Implementation, and Applications. Human Factors and Ergonomics*, pp. 1065–78. Lawrence Erlbaum, Mahwah, NJ.

Parkin, R., Marks, I. and Higgs, R. (1995) 'Development of a computerised aid for the management of anxiety in primary care', *Primary Care Psychiatry*, **1**: 115–17.

Petraub, D.P., Slater, M. and Barker, C. (2001a) 'An experiment of public speaking anxiety in response to three different types of virtual audience', *Presence: Tele-operators and Virtual Environments*, **11**: 68–78.

Petraub, D.P., Slater, M. and Barker, C. (2001b) 'An experiment on fear of public speaking in virtual reality', In Westwood, J.D., Hoffman, H.M., Robb, R.A. et al. (eds) *Medicine Meets Virtual Reality*, pp. 272–8. IOS Press, Amsterdam, Netherlands Antilles.

Proudfoot, J., Goldberg, D., Mann, A., Everitt, B., Marks, I.M. and Gray, J. (2003a) Computerised, interactive, multimedia cognitive behavioural program for anxiety and depression in general practice, *Psychological Medicine*, **33**(2): 217–27.

Proudfoot, J., Swain, S., Widmer, S., Watkins, E., Goldberg, D., Marks, I.M., Mann, A. and Gray, J.A. (2003b) 'The development and beta-test of a computer-therapy program for anxiety and depression: hurdles and preliminary outcomes', *Computers in Human Behaviour*, **19**(3): 277–89 .

Riva, G., Molinari, E. and Vincelli, F. (2001) 'Virtual reality as communicative medium between patient and therapist' In Riva, G. and Davide, F. (eds) *Communications through Virtual Technologies: Identity, Community and Technology in the Communication Age. Studies in New Technologies and Practices in Communication*, pp. 87–100. IOS Press, Amsterdam, Netherlands Antilles.

Roth, A. and Fonagy, P. (1996) *What works for whom: a critical review of psychotherapy research*. Guilford, London.

Rothbaum, B.O., Hodges, L.F., Kooper, R., Opdyke, D., Williford, J.S. and North, M. (1995a) 'Effectiveness of computer-generated (virtual reality) graded exposure in the treatment of acrophobia', *American Journal of Psychiatry*, **152**: 626–8.

Rothbaum, B.O., Hodges, L.F., Kooper, R., Opdyke, D., Williford, J.S. and North, M. (1995b) 'Virtual reality graded exposure in the treatment of acrophobia: A case report', *Behavior Therapy*, **26**: 547–54.

Rothbaum, B.O., Hodges, L., Alarcon, R., Ready, D., Shahar, F., Graap, K., Pair, J., Hebert, P., Gotz, D., Wills, B. and Baltzell, D. (1999) 'Virtual reality exposure therapy for PTSD Vietnam veterans: A case study', *Journal of Traumatic Stress*, **12**: 263–71.

Rothbaum, B.O., Hodges, L.F., Ready, D., Graap, K. and Alarcon, R.D. (2001) 'Virtual reality exposure therapy for Vietnam veterans with posttraumatic stress disorder', *Journal of Clinical Psychiatry*, **62**: 617–22.

Selmi, P.M. (1982) 'An investigation of computer-assisted cognitive behaviour therapy in the treatment of depression', *Behaviour Research Methods and Instrumentation*, **14**: 181–5.

Selmi, P.M., Klein, M.H., Greist, J.H., Sorrell, S.P. and Erdman, H.P. (1990) 'Computer-administered CBT for depression', *American Journal of Psychiatry*, **147**: 51–6.

Servan-Schreiber, D. (1986) 'Artificial intelligence and psychiatry', *Journal of Nervous and Mental Disease*, **174**: 191–202.

Shaw, S. and Marks, I.M. (1996) Computer-aided self-care for agoraphobia/panic. Paper to annual meeting of Royal College of Psychiatrists, London, July.

Stuart, S. and LaRue, S. (1996) 'Computerised cognitive therapy: the interface between man and machine', *Journal of Cognitive Psychotherapy*, **10**: 181–91.

Talley, J. (1987) *Family practitioner's guide to treating depressive illness*. Precept Press, Chicago.

Wagman, M. (1980) 'PLATO DCS: An interactive computer system for personal counselling', *Journal of Counseling Psychology*, **27**: 16–30.

Wagman, M. and Kerber, K.W. (1984) 'PLATO DCS, an interactive computer system for personal counseling: further development and evaluation' *Journal of Counseling and Clinical Psychology*, **27**: 31–9.

Weizenbaum, J. (1966) 'ELIZA – A computer program for the study of natural language communication between man and machine', Proceedings of Conference of Association for Computer Machinery, New York.

Weizenbaum, J. (1976) *Computer Power and Human Reason*. WH Freeman, San Francisco.

Wiederhold, B.K. and Wiederhold, M.D. (2000) 'Lessons learned from 600 virtual reality sessions', *CyberPsychology and Behavior*, **3**: 393–400.

Wright, J.H. and Wright, A.S. (1997) 'Computer-assisted psychotherapy', *Journal of Psychotherapy Practice and Research*, **6**: 315–29.

Wright, J.H., Salmon, P., Wright, A.S. and Beck, A.T. (1995) 'Cognitive therapy: a multimedia learning program', presented at the American Psychiatric Association annual meeting, Miami Beach, FL, May.

Wright, J.H., Wright, A.S. and Beck, A.T. (2002a) *Good Days Ahead: The Multimedia Program for Cognitive Therapy*. Mindstreet, Louisville.

Wright, J.H., Wright, A.S., Basco, M.R., Albano, A.M., Raffield, T., Goldsmith, J. and Steiner, P. (2001) 'Controlled trial of computer-assisted cognitive therapy for depression', World Congress of Cognitive Therapy, Vancouver, Canada, July.

Wright, J.H., Wright, A.S., Salmon, P., Beck, A.T., Kuykendall, J., Goldsmith, J. and Zickel, M.B. (2002b) 'Development and initial testing of a multimedia program for computer-assisted cognitive therapy', *American Journal of Psychotherapy*, **56**: 76–86.

9 The computer plays therapist: the challenges and opportunities of psychotherapeutic software

KATE CAVANAGH, DAVID A. SHAPIRO
AND JASON S. ZACK

The development of, and research into, computerized applications in psychotherapy is reviewed in the previous chapter, and elsewhere in this volume. The ethical and logistical considerations regarding the implementation of psychotherapeutic software have also attracted an abundant literature. In this chapter we address the challenges and opportunities of systems designed to assist or replace the therapist in the delivery of psychotherapeutic work: those programs where the computer application plays therapist. The limitations of a single chapter necessitate the exclusion of an in-depth discussion of computer software designed for clinical training or assessment, although assessment issues regarding client suitability for computerized psychotherapy are discussed.

First we offer a working definition of computerized psychotherapy and the implications of these applications in the context of community psychotherapy. Then we address core ethical issues in the use of computerized psychotherapy: is it safe? is it second best? can it be trusted? Finally, we consider the logistics of implementing psychotherapeutic software practice to marry the needs of clients, practitioners and service frameworks.

WHAT IS COMPUTERIZED PSYCHOTHERAPY?

Psychotherapeutic software can reduce client–practitioner contact time, meaningfully and effectively deliver the training elements of the psychotherapeutic intervention and transmit positive non-specific ingredients of therapy such as empathy and motivation in the absence of human interaction. Thus one way to consider therapeutic software is along these three dimensions of time, training and relationship.

Client–software contact time

To date, the term 'computerized psychotherapy' has represented a broad range of programs which might assist or replace the human practitioner. An application is computerized when some or all of the therapeutic ingredients are offered by a computer program and via a computer interface (PC, laptop, palmtop, interactive voice response, mobile telephone). The role of computerized psychotherapy in client care might be as a sole resource, an adjunct or a sequential ingredient of a wider therapeutic pathway. Client–software contact time ranges from minimal applications where just 5 per cent of psychotherapeutic interaction is between client and software (Hassan, 1992), through moderate applications designed to reduce client–therapist interaction by half (for example Wright et al., 2002) to programs where the vast majority of the psychotherapeutic interaction takes place between the client and software package (for example Kenwright et al., 2001; Proudfoot et al., 2003a).

Where minimal contact psychotherapeutic software may be viewed simply as an adjunct to traditional psychotherapeutic work, more ambitious programs have the potential to seriously alter the nature of the therapy process. In the case of *FearFighter*, use of the program reduces client–practitioner contact time by 86 per cent compared with comparable face-to-face intervention (Kenwright et al., 2001). In the case of *Beating the Blues*, the program has been designed such that following assessment, no clinician input other than progress-monitoring is required for most clients. In both cases, the software guides clients through many or all the processes involved in the traditional therapeutic setting: pinpointing specific problems and therapeutic goals; problem and mood-monitoring; training techniques to manage specific problems; target-setting for the implementation of these techniques in day-to-day life; and coping with setbacks to progress. Not only do these programs eliminate much of the need for traditional client–practitioner interaction in terms of contact time, but they also deliver the core ingredients of psychotherapy as it is classically defined.

Client–software therapeutic training

Software can also be developed to deliver training in therapeutic techniques or knowledge. This kind of application is consistent with Meltzoff and Kornreich's (1970, p. 6) definition of psychotherapy itself:

> The informed and planful application of techniques derived from established psychological principles, by persons qualified through training and experience to understand these principles and to apply these techniques with the intention of assisting individuals to modify such personal characteristics as feelings, values, attitudes and behaviours which are judged by the therapist to be maladaptive or maladjustive.

In this context, computerized psychotherapeutic tools can be viewed as one resource that offers personalized guidance, information and therapeutic strategies to help clients understand and alter unwanted feelings, thoughts and behaviours. Inherent to this definition is the empowerment of the client. The client is encouraged to identify problems and unwanted actions, to claim primary agency in the psychotherapeutic process and to utilize available resources to accomplish desired outcomes.

To be recognized as a legitimate form of psychotherapy, this kind of self-management requires a point-of-view shift within the psychotherapeutic community. One barrier to the acceptance of computerized psychotherapy is the high value placed on the client–therapist relationship and the reproduction of this alliance with a computerized practitioner.

The changing face of psychotherapy: the client–software relationship

A cornerstone of nearly every brand of psychotherapy is that the client–therapist relationship is of primary importance (Gelso and Hayes, 1998; Wampold, 2001).

Reviews of psychotherapy research (for example Horvath and Luborsky, 1993; Orlinsky et al., 1994) have repeatedly identified the 'therapeutic alliance' as the most powerful correlate of therapeutic change and it appears that the therapeutic relationship is responsible for much of the benefit associated with psychotherapy.

Since by definition a computer program cannot form a person-to-person relationship with a client, this might be taken to imply that computerized psychotherapy cannot possibly be effective. Spero (1978, p. 279) has argued that although computers

> may be 'intellectually' capable of performing certain skills that have up to now been wholly associated with the human mind, still other *affective responses*, such as empathy, sympathy and compathy, are not within the computerised therapy device's ken.

However, early studies of automated psychotherapeutic outcomes revealed that non-human delivery of therapeutic ingredients may actually exceed their human counterparts in certain circumstances. For example, Lang et al.'s (1970) study of DAD, an automated systematic desensitization program for snake phobia, revealed better outcomes on some measures for DAD than therapist-guided treatment, as noted in Chapter 8.

A meta-analytic review by Peck (2002) compared psychotherapies delivered by face-to-face and non-face-to-face methods. With only a few exceptions, the effects of non-face-to-face methods were equivalent or more powerful than face-to-face techniques. Furthermore, Peck suggested that some elements of the therapeutic alliance such as agreement on task and treatment goals may be beneficial and achievable using written or computerized materials, whereas actual contact between practitioner and client carries the potential for negative effects due to the inherent difficulty in disclosing highly sensitive or embarrassing personal information, as well as the risk of exploitative therapist behaviour (for example Pope, 1990).

We believe that it would be unwise to reject computerized therapy a priori on the basis of a definition of psychotherapy that requires interpersonal contact between therapist and client: a restrictive semantic quibble. The development of interactive multimedia software further challenges these assumptions. Psychotherapeutic software can be designed to maximize user engagement by conveying empathy, warmth and understanding without direct encounter with a human (Marks et al., 1998).

We conceptualize therapeutic change as client accomplishment, enabled by a variety of potential resources. There are strong ethical, political and empirical reasons for choosing this conceptualization over a traditional, alternative formulation viewing the practitioner as the change agent, performing manipulations upon a passive client.

Ethical and political considerations relate to the primacy accorded to the client or consumer of professional services as a valued individual to whom professional mental health practitioners owe respect and care (Lovell and Richards, 2000; Mead and Bower, 2000). Empirically, research grounded in the drug metaphor (therapy as a prescribed and dispensed intervention) has proved disappointing, in that few 'active ingredients' of psychotherapy have been proven to account for therapeutic change (Stiles and Shapiro, 1989).

Our view of change as an accomplishment of the client invites us to locate most, if not all, the psychological processes involved in therapeutic change within the client. From this perspective, a positive therapeutic alliance enables such client processes as hope (Snyder et al., 1999) and self-efficacy. There is no reason why such processes cannot be facilitated by the use of non-human resources such as self-help books and psychotherapeutic software. From a cognitive-behavioural therapy (CBT) perspective, a positive therapeutic relationship may provide reinforcement of change efforts (Beck et al., 1990), or incentives for clients to remain engaged in therapy (McNeil et al., 1987; Waddington, 2002). But again, face-to-face contact with another person is not the only means of securing these outcomes. Interpersonal learning can take place vicariously through books or computer programs – virtual representations of real or fictional people from whose experience clients can learn how best to tackle their own problems.

In sum, we believe that the therapeutic alliance literature offers no convincing reason to deny the possibility that computerized therapy can help clients to achieve the gains they might otherwise receive from face-to-face therapy. Once we identify the psychotherapeutic outcome as primarily a client accomplishment, then there is no longer the need to insist upon a two-human dyadic relationship. A consequent definition of psychotherapy can include self-management approaches including computerized applications and the use of psychotherapeutic software.

ETHICAL ISSUES IN THE USE OF PSYCHOTHERAPEUTIC SOFTWARE

As with the implementation of any healthcare service, the application of psychotherapeutic software might raise a number of ethical concerns. First, we must consider whether such applications are safe and able to manage crisis situations effectively. Second, we must consider whether these programs represent a 'second best' to clients and practitioners. Finally, we must consider issues of client trust and confidentiality: are these systems misleading and are they robust when it comes to data storage?

First do no harm: is psychotherapeutic software safe?

One of the most common concerns about computer-mediated psychotherapy is the ability (or inability) to attend to crisis situations. People suffering from mental health problems who can benefit from psychotherapy are also a group who, on average, may be at increased risk of suicide and self-harm. Consider, for example the population of depressed clients who might use *Beating the Blues* or *MindStreet.* Diagnosable depressive disorders are implicated in between 40 and 60 per cent of instances of suicide (Fowler et al., 1986; Clark and Fawcett, 1992; Henriksson et al., 1993) and it has been estimated that between 10 and 15 per cent of people diagnosed with major depressive disorder eventually kill themselves (Clark and Fawcett, 1992; Maris et al., 1992), a risk rate 20 times higher than population-expected values (Harris and Barraclough, 1997).

If one of the psychotherapist's duties is to identify and manage patient risk, is safety compromised by the use of psychotherapeutic software? In a traditional face-to-face psychotherapy setting, the practitioner has an ethical (and often legal) responsibility to ensure the client's safety or the safety of other individuals against whom an aggressive client has threatened harm. The question for the present discussion is whether or not offering computerized therapy applications (and/or their developers) carries a similar ethical responsibility.

Where psychotherapeutic software is used adjunctively with face-to-face therapist contact, risk assessment and monitoring functions are not typically performed by the software. For example, in the delivery of *FearFighter*, and other computer-delivered therapies (*COPE, BALANCE*, BT *STEPS*) at the London Stress Self-Help Clinic, patients are prescreened and those at risk are not offered a computerized intervention. In the case of *MindStreet*, where clients see a practitioner as well as the computer during each session, risk-monitoring remains in the hands of the human not the program.

Where psychotherapeutic software is largely a 'stand-alone' intervention, risk assessment, monitoring and responsibility may be more complex and sophisticated. Therapists are appropriately sensitive to risk issues and may therefore be wary of computerized therapy on the grounds that clients need to be monitored for risks such as suicide. The individual or organization that offers computerized therapy may reasonably have a duty to prevent suicide, or any serious harm to self or others. The delivery of computerized therapy must therefore include a reliable system for the detection and appropriate communication of risk. Risk management is implicated in assessments of both client suitability and client progress.

Risk assessment: the suitability of the program

In the case of stand-alone psychotherapeutic software, risk assessment prior to the program is the practitioner's responsibility. Referral is managed by a health professional (usually the patient's family doctor or mental health worker), who must assess the patient's suitability for the program. Referring staff are advised that the program is not suitable for those contemplating suicide. For example, support staff working with *Beating the Blues* participate in training sessions and receive printed manuals in which this exclusion is reinforced.

Risk-monitoring: session by session

Risk status can change rapidly and the initial assessment or screening may not be sufficient to detect dynamic safety issues. Where software is delivered independently, session-by-session monitoring is recommended. For example, *Beating the Blues* monitors patients' thoughts and plans of suicide at the beginning of each weekly session. Examples and case example interaction from the program can be seen below.

Risk-monitoring in *Beating the Blues*

Session 1: risk-monitoring (repeated at the beginning of each session)

Computer (voice-over, accompanied by screen graphic): Your doctor will also get a brief progress report each week. It will contain your anxiety and depression ratings, how distressing your problems have been and whether you have had any upsets or thoughts of suicide.

Computer (voice-over, accompanied by screen graphic and text): I need to ask you each week whether you have had any thoughts of suicide. Have you had any thoughts of suicide in the last week?

1. Yes
2. No

Patient: No

Screenshot 9.1 Thoughts of suicide

Where no suicidal ideation is indicated, the program moves on to a discussion of recent upsets and disappointments. Where the patient reports thoughts of suicide (response: yes), the program continues thus:

Computer (voice-over, screen text and graphic with multiple choice responses): How often have you thought about ending your life in the last week?

1. Once
2. Twice
3. Three times
4. More than three times

Patient: any response

Screenshot 9.2 Frequency of suicidal ideation

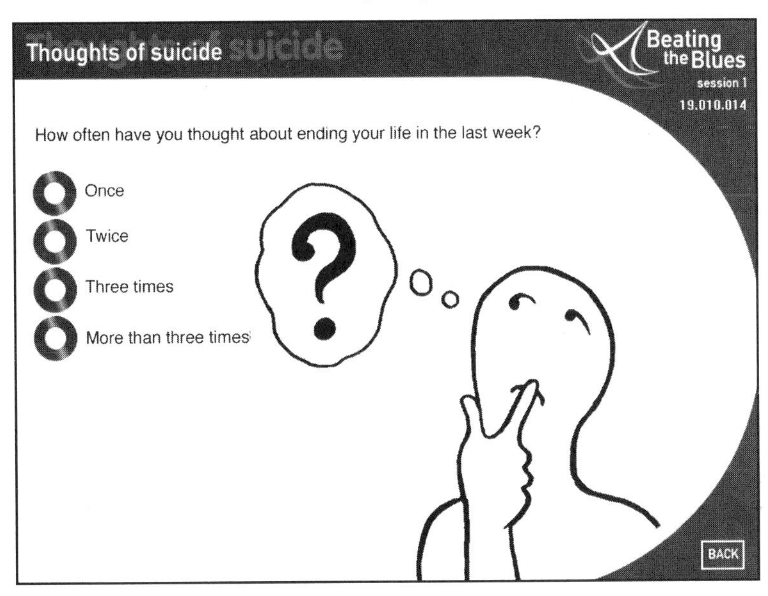

Computer (voice-over, screen text and graphic with multiple choice responses): How seriously have you planned to carry it out?

Screenshot 9.3 Suicidal intention

0 1 2 3 4 5 6 7 8

not very seriously very seriously

Patient: response 0–4

Computer (voice-over and screen graphic): I'm sorry you've been thinking about suicide. Things have obviously been very bad for you. If you find your plans are getting any more serious please stop using *Beating the Blues* and see your doctor or someone else who can help you.

Patient: response 5–8

Computer (voice-over and screen graphic): I'm concerned you've been thinking about suicide. Things have obviously been very bad for you. Please see your doctor or someone else who can help you straight away.

Screenshot 9.4 Crisis management

The client's responses to the risk interview are indicated in full on his or her progress report. Where active suicidal ideation is recorded this is highlighted on the progress report for the attention of his or her clinician.

The limitations of this monitoring process lie in the establishment of a viable and effective 'safety net' system such that risk information is appropriately communicated to a clinician or other person in accord with local practice standards. For example, when delivered in primary care environments, the client's progress report can be delivered to and checked by a clinician before he or she leaves the building.

Possible barriers to the effective implementation of such a patient-monitoring system include training and the resource constraints of delivering psychotherapy in primary care. It is imperative that the clinical helper (typically a nurse or administrative member of staff who supports the day-to-day running of the program) and the clinically responsible party (for example family doctor or therapist) understand the progress report information and make time to read and sign off the progress report each week. Furthermore, clinical helpers need training and support, or they may feel uncomfortable or ill equipped to deal with the risk information provided to them.

Future programs might overcome these barriers by linking the therapeutic software to the local case management database, such that progress reports can be delivered instantaneously to the clinician's desk or a case management call centre, marked for urgency as appropriate to the client's level of risk. Again, local safety protocols must be applied where risk is identified.

Risk-monitoring: covert risk cues

It is possible that a client may not report suicidal intent, but their presenting problems may contain a risk cue, such as an intention to harm themselves or other people. At present the majority of psychotherapeutic software in use does not employ natural language parsing. Client problems are simply repeated back by the computer verbatim. Future programs could be designed to actively detect words and phrases such as, '...kill myself', 'life isn't worth living', and 'I swear I'm going to make him pay...' to name but a few examples. In the case of such phrase detection, risk alerts, similar to those activated in overt risk-monitoring, could be presented to the clinician. However, such monitoring may contravene established confidentiality agreements on program uptake. In the case of *Beating the Blues*, users are informed that although their clinician will have access to risk information, mood and problem-monitoring, all other inputs to the program will remain confidential. The implications of a caveat permitting access to phrases that activate a risk indicator are unknown.

Risk-monitoring: homicide risk

To our knowledge no current computerized psychotherapy program is designed to assess, detect or direct cases of other harm, homicidal ideation or intent, although such situations could presumably be handled in a similar manner to situations involving self-harm.

Safe, safer, safest

Despite vocal professional concern, there is little evidence to suggest that computers do not present a safe delivery vehicle for psychotherapeutic intervention. Indeed, psychotherapeutic software may offer better risk detection than a human psychotherapist. The computer interface may have a disinhibiting effect (Joinson, 1998), allowing clients to be more honest about their feelings and intentions. Thus, clients are quite willing to undertake computerized assessment, and are more likely to report sensitive information, including those associated with suicide, to a computer than to a human practitioner (Erdman et al., 1985). Moreover, computerized risk assessments have been demonstrated to be more accurate in predicting suicide attempts than human therapists given the same information (Greist et al., 1973).

Is psychotherapeutic software a second best?

Effectiveness

The outcome studies available, although limited in number, indicate that psychotherapeutic software can benefit people in need (see Chapter 8). In the case of programs designed for the management of anxiety and depression using cognitive-behavioural techniques, treatment effect sizes are comparable to those found in studies where the same techniques are delivered in the traditional face-to-face manner (Ghosh and Marks, 1987; Selmi et al., 1990; Kenwright et al., 2001; Wright et al., 2002; Proudfoot et al., 2003a and b). No psychotherapy is a panacea but, in terms of clinical outcomes, there is no reason to believe that well-designed programs, offered to appropriate clients, are a second best to traditional psychotherapy.

Client acceptance and satisfaction

Acceptability and satisfaction with psychotherapeutic interaction is not measured solely by changes in clinical outcome measures. Client utilization, continuation and preference for any healthcare service will be determined by a range of factors including delivery mode, motivation, continuation benefits (for example improved wellbeing) and discontinuation benefits (for example what is gained from switching formats).

The embrace of psychotherapeutic software by both practitioners and clients will be driven, at least in part, by the acceptability of self-help approaches in general and computerized delivery mechanisms in particular. Self-help approaches to mental health concerns are gaining in popularity. More than 2,000 self-help books are published each year in the United States (Norcross et al., 2000), and self-help Internet searches and group work gain in popularity year on year.

The acceptance of psychotherapeutic software reflects this trend. A survey of potential users of self-help psychotherapies in the UK found that 91 per cent of respondents wanted to access self-help psychotherapies via a computer system and that accessing these systems by CD-ROM at home, in the GP surgery or in a local community resource centre were amongst the most popular routes (Graham et al., 2001). These findings suggest that people suffering mental health problems who are typically offered psychotherapy do find psychotherapeutic software an acceptable alternative.

People using psychotherapeutic software also report finding the programs interesting, helpful and recommendable (Wright et al., 2002; Proudfoot et al., 2003a). These findings support the claim that they are not considered second best by users. Indeed, as already noted above, psychotherapeutic software may be more acceptable than face-to-face methods for some clients. Where problems are of a sensitive nature, or forming trusting relationships is a challenge, computerized psychotherapies might offer a preferable alternative. Wagman (1980) found that a substantial minority of people using the computerized *Dilemma Counselling System* (42 per cent) felt more at ease with the computer than when they saw a counsellor. In a beta test of *Beating the Blues*, 90 per cent of participants felt that the computerized psychotherapy was as good as, or better than, previous treatments including medication and face-to-face counselling or psychotherapy (Proudfoot et al., 2003a).

A further measure of treatment acceptability and satisfaction is that of treatment continuation. Discontinuation rates for psychotherapeutic software are reported to be a little over one-third (Kenwright et al., 2001; Proudfoot et al., 2003a), a figure echoing that for face-to-face CBT (for example Watkins and Williams, 1998). Again, these figures give no reason to suggest that clients find psychotherapeutic software a second best to traditional psychotherapy.

Therapist acceptance and satisfaction

For computerized psychotherapies to be established within healthcare systems, it is imperative that local practitioners, including psychotherapists, find these delivery methods acceptable. In presenting *Beating the Blues* to UK professional audiences over some three years, we have discerned a substantial shift in attitudes towards the program. Initially, there was a considerable amount of scepticism or resistance to the very idea that a computer program could deliver CBT. Some CBT therapists feared that computerized CBT might reinforce unhelpful caricatures of CBT as a simple-minded, mechanistic approach lacking the human sensitivity of rival approaches to psychological therapy. Others acknowledged a concern that, if successful, this approach could undermine the case for increased provision of face-to-face therapy by presenting a viable and lower cost alternative.

However, as would be expected from their grounding in an evidence-based approach, CBT therapists' attitudes have shifted with the availability of research data supporting the efficacy of *Beating the Blues* and colleagues' reports of clients finding the program helpful. In addition, a growing awareness of the *stepped care* approach (Haaga, 2000; Lovell and Richards, 2000), in which limited resources are carefully targeted according to the client's assessed level of need, allows room for the full range of delivery methods as appropriate to different clients. This largely neutralizes any perceived threat to face-to-face CBT provision. It also implies a significant role for computerized therapy in enabling therapists to concentrate their efforts on increasing client access to services and work with clients whose need for face-to-face therapist contact is greatest, with some reduction in the pressure of sheer numbers awaiting treatment.

Second best to what?

Despite a growing body of evidence that psychotherapeutic software can offer an effective and acceptable alternative to traditional psychotherapies, there may still be some doubt amongst practitioners that a computer program can really offer an equivalent experience to the traditional psychotherapeutic process. Indeed, these reservations may be well-founded. Given the choice between psychotherapy sessions with a computer and a person, it seems unlikely that more than a minority would choose the psychotherapeutic software. However, it must be noted that this is rarely a viable set of alternatives. Psychotherapeutic software offers an alternative not to an abundance of open face-to-face psychotherapy sessions, but to months (or sometimes even years) on waiting lists, high thresholds for entry into therapy services and a consequent vast unmet population need for care. Other debates aside, psychotherapeutic software certainly offers a powerful, proven alternative to nothing and to the iatrogenic effects of unmet needs for early intervention for people with mental health problems. Ultimately, the question of equivalence is really not nearly so important as whether computerized therapy is effective and a satisfactory treatment method compared to nothing at all, or treatment as usual (waiting lists and so on).

Trust and confidentiality

Issues regarding client trust and confidentiality in using psychotherapeutic software overlap the boundary between a discussion of ethical and logistic issues. Protocols to ensure data confidentiality can only be effective where they are adhered to in practice. This section explores trust and confidentiality issues relevant to the use of psychotherapeutic software including client identity and data storage/access and protocols.

Client trust

Whether a client enters a psychotherapeutic encounter with a human or a computer, trust issues are involved. Can I trust that this will do me no harm? Can I trust that this encounter is in my best interests? Can I trust that my identity and the information that I disclose about my problems and myself will be treated appropriately? The first two questions here converge with issues regarding perceived efficacy (discussed above) and therapeutic accreditation to be discussed later in the chapter, the latter regards client confidentiality and data protection in the healthcare setting.

Identity

Where psychotherapeutic software is used in a healthcare setting, client identity is available to service staff and concurrent service users, via the usual routes. Staff will have access to local appointments including those to work on psychotherapeutic software and concurrent service users may identify program users entering or leaving the therapeutic setting. However, it has been suggested that the stigma associated with seeking help for mental health problems may be reduced in the case of psychotherapeutic software (for example Marks et al., 1998). Clients may be seen as working to help themselves, working on the computer or learning new skills, rather than 'seeing a shrink'.

Where psychotherapeutic software is delivered remotely, client identity may be concealed. This may offer some advantages in making services more accessible to some groups previously reluctant to seek help. Working privately may offer an attractive option to some people who have difficulty in forming relationships, where problems are of a sensitive nature or for whom admitting to a mental health problem may have social or work-related implications.

Data access and data protection

In developing psychotherapeutic software, consideration to appropriate levels of data access and data protection must be employed. Entry to the client's therapeutic sessions should be password-protected, the password being generated and held solely by the client.

Healthcare professionals holding clinical responsibility for client well-being may wish to have access to progress information regarding mood and problem-monitoring, as well as risk information (discussed above). The parameters of this data access must be made clear to the client. In the case of *Beating the Blues* and *MindStreet*, clinicians and clients can access duplicate copies of a session-by-session progress report charting such cumulative information. But data entered into the system by the client during the rest of the therapeutic encounter is not accessible to the clinician. For example, in *Beating the Blues*, clients enter items on a fear hierarchy or responses to the 'four challenging questions' regarding thinking errors. Such information is encrypted and retained within the program database for later retrieval by the clients but not, without permission, their clinician. The confidential features of psychotherapeutic software may be viewed as an advantage by some clients who wish to work through their problems privately.

LOGISTICAL CONSIDERATIONS FOR PSYCHOTHERAPEUTIC SOFTWARE

The implementation of psychotherapeutic software services offers both solutions and challenges. Such programs can make psychotherapeutic sessions of a consistent quality available for more clients, in more places, more of the time. Moreover, psychotherapeutic software may offer a number of practical advantages to researchers and practitioners as well as to healthcare services in attempts to meet needs for care regarding common mental health problems typically seen in the psychotherapeutic environment.

This section outlines the practical advantages of implementing psychotherapeutic software programs and explores the barriers and drawbacks to their successful integration into pathways of care.

Accessibility

Common mental health problems such as anxiety and depression have extremely high incidence. At any one time about 13 per cent of the British population suffer from anxiety states or depression with combinations of the two being very common (Melzer et al., 1995; Goldberg and Huxley, 1997), and similar figures are seen internationally. The largest epidemiological study of depression and related disorders in the UK indicated that 1 in 6 adults aged 16–64 had a neurotic disorder. This represents more than 7 million people suffering from anxiety and or depression in the UK at any one time. In any given one-year period, 9.5 per cent of the population, or about 18.8 million American adults, suffer from a depressive illness and 13 per cent, or about 19.1 million people, suffer from an anxiety disorder (Robins and Regier, 1990; National Institute for Mental Health, 1998), while in Australia in any one year 6 per cent of the population suffer from major depression and 15–30 per cent experience an anxiety disorder during their lifetime (Australian Institute of Health & Welfare, 2002).

At present between just 12.5 and 14 per cent of those currently suffering from anxiety or depression are receiving treatment for them (Ohayon et al., 1999; Bebbington et al., 2000a and b), and in the UK just 8 per cent of those are receiving treatment which includes psychotherapy or counselling (Bebbington et al., 2000a). This means that of 7 million people in the UK alone who might benefit from psychotherapeutic work for problems of anxiety and depression each year, just 70,000 do (Gournay, cited in Lovell and Richards, 2000). This represents a hundred-fold shortage of services. This paucity of psychotherapeutic treatments largely reflects the lack of appropriately trained practitioners. For example, there are just 585 British Association for Behavioural and Cognitive Therapies (BABCT) accredited CBT therapists in the UK at present (Lomas, personal communication). To meet population needs, entry to training programmes would need to be increased by an unattainable magnitude. Furthermore, the geographic distribution of trained psychotherapeutic practitioners is inequitable (Shapiro and Cavanagh, 2002). Such scarcity and inequity in CBT is no doubt mirrored both internationally and in other schools of psychotherapeutic method.

Psychotherapeutic software offers a solution to this shortfall. Such programs can rapidly increase access to help for populations currently in need. Psychotherapeutic software packages that permit an annual throughput of up to 200 clients may be a welcome addition to local services struggling to meet the needs of their caseload. Not only can psychotherapeutic software be equitably distributed but, where service models permit, they can be accessed both during the working day and out of hours, offering increased access to care. However, it must be recognized that much of the burden of increased client volumes will fall to administrative rather than therapeutic staff. Administrative practices will need to be adjusted to accommodate the changing nature of the service.

The flexible nature of psychotherapeutic software means that it can be adapted to meet client needs in ways that a human practitioner cannot. For example, psychotherapeutic software can be engineered to maximize client–therapist match. Above a certain threshold, programs can be attuned to a client's abilities in terms of reading, writing or drawing. Voice-overs, case studies and examples can be designed to offer interactivity with a system suited to the client's specific preferences, problems, demographic characteristics, first language and so on. At present such flexibility is not typical of available psychotherapeutic software. However, with increased recognition and funding, these features will be made available.

Client suitability

Zarr (1984) proposed treatment selection guidelines for computer-mediated psychotherapy, based on Frances and Clarkin's (1981) guidelines for treatment selection in psychiatry. Zarr suggested that we must consider the following criteria, which can equally be applied to psychotherapeutic software today:

1. treatment format
2. treatment orientation
3. treatment setting, duration and frequency
4. therapist–patient match
5. treatment with or without medication
6. indications for prescribing no treatment.

Zarr argued that 'where psychoanalytically orientated psychotherapy is indicated, computer mediated psychotherapy is not, at present, an adequate substitute' (pp. 58–9). For other individual therapy psychotherapeutic indications, he suggested that if the treatment orientation was available via computerized delivery, and computerized methods were appropriate in terms of setting, duration and frequency, and the client was appropriately skilled and motivated to cooperate with the psychotherapeutic software, then such application may be considered a treatment option.

In addition to behavioural and cognitive psychotherapies, Zarr (1984) argued that psychotherapeutic software may be useful for guidance and vocational counselling (see Wagman, 1980; Wagman and Kerber, 1984). Psychotherapeutic software may offer an efficient program where longer therapy is indicated, but some specific problems can be managed by the computer, working in conjunction with face-to-face therapy for more complex problems. Nor is there any evidence that computer therapy is contraindicated in inpatient settings.

However, even where a therapeutic approach available via psychotherapeutic software is indicated in a client care pathway, the question of which clients can benefit from this delivery model is still in its infancy. Little is known about the differential client benefits of psychotherapeutic software.

New case pathways

Computerized psychotherapies offer a range of new possibilities for case management. The emergence of effective self-help programs, including psychotherapeutic software, offers new steps in a 'stepped care' approach (Haaga, 2000; Lovell and Richards, 2000). Moreover, psychotherapeutic software may be offered sequentially within a defined care pathway. Rather than waiting for 18 months to enter face-to-face psychotherapy, clients can gain immediate access to psychoeducational and problem-focused psychotherapeutic software in a sequential manner. Such access may curtail the iatrogenic process of unmanaged mental health problems, but also promote learning which may reduce the number of face-to-face sessions necessary for agreed therapeutic closure. Alternatively, psychotherapeutic software might be offered as a relapse prevention module in the care pathway. In either case, psychotherapeutic software might offer a number of ways of augmenting care pathways to benefit both the user and clinician.

Consistent and contemporary psychotherapeutic methods

A further practical advantage of psychotherapeutic software is the delivery of consistent and contemporary psychotherapeutic methods and techniques. Unlike human therapists who may vary in their approach between sessions and between clients, psychotherapeutic software offers a reliable translation of proven techniques, which can be updated to reflect the dynamic evidence base, independently of extraneous factors. As Colby et al. (1979) noted, there are a number of ways in which the computer might improve upon the unreliable human therapist:

1. computers are polite, friendly and always have good manners
2. computers do not get tired, bored or forgetful
3. computer programs never get irritated, annoyed or reproachful of the patient and will not show facial expressions of shock, contempt or surprise
4. computers do not have family problems
5. computers never get sick or hungover
6. computers never try to sleep with the patient!

In addition to overcoming human foibles, psychotherapeutic software may be more quickly and cost-effectively 'retrained' than the human psychotherapist. The translation of the psychotherapeutic process into psychotherapeutic software is both limited and permitted by our knowledge of what works in traditional psychotherapy. As Marks et al. (1998, p. 164) state:

> we can only construct an effective computer system to help a given problem when we know exactly which of a clinician's countless actions help that problem. If we aren't crystal clear what the clinician does that is useful for that problem then we can't build a relevant system. As we learn more about which therapeutic ingredients help what, so our ability to reproduce them by computer can evolve.

Success in developing effective psychotherapeutic software relies on the ability to articulate the decision pathways relevant to a proven technique. Once these can be articulated psychotherapeutic software can be developed and updated to incorporate our unfolding knowledge of what works for whom.

Cost-effectiveness issues

This discussion of the opportunities afforded by psychotherapeutic software clearly leads to an analysis of the issue of cost-effectiveness. Whilst development costs are high (it is estimated that the state-of-the-art multimedia packages for the treatment of anxiety and depression have each to date entailed development costs in the region of several million dollars, (see Marks et al., 1998), these figures are likely to decrease as our knowledge base increases and technological advances permit.

The implementation costs of psychotherapeutic software can be understood in terms of several components including the manufacturer's price for such products, the direct and indirect costs of staffing to manage these services and the capital overheads for locating these services within healthcare facilities. Where psychotherapeutic software is conceived to replace human psychotherapists, absolute costs would no doubt reflect a saving. However, this is rarely the intention. Most psychotherapeutic software is designed to augment rather than replace human services. By increasing the availability of psychotherapy, psychotherapeutic software is unlikely to offer an absolute cost saving to local services. Rather, its intention is to offer a cost-effective strategy for increasing availability of psychotherapy where appropriately trained human therapists cannot meet a population need for care.

Training and implementation considerations: technical and psychological support required by users, clinical helpers and referrers

The implementation of psychotherapeutic software packages to manage mental health problems requires technical and psychological training and support for three parties: the user, the clinical helper and the referrer. The user must be made at ease to work with the program interface and understand the contact parameters of the psychotherapy. The clinical helper must be trained to provide necessary support for the user in getting started and progressing through the program. The clinical helper must also understand and implement local protocols regarding progress-monitoring and risk assessment. The referring party (for example a family doctor or therapist) must understand the possibilities and limitations of the psychotherapeutic software program. They must understand for which problems the program is indicated, which clients are suitable for this mode of therapeutic delivery and their own role in supporting the clinical helper and monitoring the user's progress.

All these requirements represent a training need within the healthcare setting. Who is equipped to deliver such programs, and who is equipped to supervise trainees? As the use of psychotherapeutic software becomes a healthcare standard, such processes must be formalized. Here professional concerns mirror those in the use of other self-help materials. The majority of psychologists and psychotherapists recommend self-help books and other resources (groups, websites, audio tapes, movies) to clients (Norcross et al., 2000) and yet very little training is received by professionals in their use.

Problem-based treatment indications are likely to mirror those in traditional psychotherapy and treatment of choice guidelines are available internationally (Roth and Fonagy, 1996; Department of Health, 2001). However, client suitability guidelines for psychotherapeutic software are still in their infancy (see client suitability section above).

For psychotherapeutic software to be meaningfully implemented it much reach agreed standards of technical reliability. No computer package (even from the largest manufacturers) can guarantee error-free functioning, but future professional standards for psychotherapeutic software ought to agree what proportion of errors is permissible in practice and accredit programs accordingly. Where technical errors do occur, service standards must be dictated such that they are rapidly rectified. Error severity must be easily classified by defined troubleshooting protocols. Wherever possible, clinical staff should be trained to remedy simple software and hardware faults – for example where restarting the program can overcome system errors. In the case of more significant technical difficulties, manufacturers must be willing and able to offer rapid solutions, for example a replacement system, whilst difficulties are identified and overcome.

Business issues

How clients pay for regular therapy has a great deal to do with practice and process. Thus it is important to consider how clients pay for psychotherapeutic software. We expect to see a variety of pricing/packaging options. For example, self-contained counselling programs may be pricey because they use advanced technology and are seen as medical tools. Conversely, they may be priced inexpensively if developers want them to be accepted into the self-help marketplace, which is used to buying cheap paperback guidance. The fee structure will largely determine the client base and typical use of the software.

Third-party insurance reimbursement is a critical component of mental healthcare in the US and elsewhere. Although insurance companies do not generally pay for self-help books, they will often cover medical equipment that is prescribed by a physician. If a psychiatrist, for example, were to prescribe the psychotherapeutic software along with medication to treat depression, it might be considered to be an insurable benefit. Insurance companies might be especially willing to cover the software if empirical research could show that software reduced the utilization of face-to-face services. If computerized therapy could be proven to have processes and outcomes similar to traditional face-to-face therapy, perhaps a case could even be made that computer-facilitated sessions could be reimbursed in the same way as individual psychotherapy.

Licensing issues

Another interesting ethical and logistic question centres on licensing and regulatory issues. Therapists are required to have some sort of professional certification, accreditation or licence by the jurisdiction within which they practice. Should some sort of oversight board regulate computerized therapy as well? Will healthcare professionals need a licence in order to offer psychotherapeutic software? Will individual software packages require accreditation for use?

Individual treatment approaches (for example cognitive-behavioural therapy) normally do not require certification but it seems different if the computer is truly acting as the therapist, not merely being another intervention that is itself delivered by a therapist. If the computer acts as therapist, we may need to consider whether such programs need to be regulated or certified by a central agency that is responsible for ensuring that healthcare professionals are adequately competent to provide services. As technology becomes more sophisticated and computers become more person-like, this will be a more important issue. Another concern arises with remote-access programs and software that is distributed across jurisdictional boundaries. For example, each US state and each European country is individually responsible for licensing the practice of mental health professionals. Would a therapy software developer need to get the software approved by the Board of Psychology in each of the 50 states in America? What about Canada, the UK and other countries where the software is intended to be distributed? These are questions that do not yet have definitive answers but which will need to be addressed in the near future.

'This is not a therapy'

At present there is no guidance or regulation regarding the use of psychotherapeutic software in mental health practice. Psychotherapeutic software is still such an unfamiliar concept that even its authors are not sure how to present it to the world. Disclaimers abound, for example on an opening screen, the *Dream Toolbox* (Zack and Hill, 2001) states: 'this computer program should not be considered as any kind of professional counselling or psychotherapy. It is merely a fun, informative, and potentially useful method ...'

From the perspective of face-to-face therapy practitioners whose names appear on registers governed by strict codes of conduct, such disclaimers may seem essential. However, from the perspective of the huge global market in self-help books, most of whose dust jackets make strong claims of potential benefits to readers, such disclaimers appear overcautious. There is no evidence to justify the much greater caution surrounding the presentation of computerized therapy as compared with self-help books. This caution may serve only to undermine the user's confidence in the benefits of using the product and thereby hamper development of the requisite degree of therapeutic optimism. Software accreditation or local practice licence to offer psychotherapeutic software may help to overcome such caution.

RESEARCH BENEFITS OF FULLY SPECIFIED INTERVENTION

A final practical advantage of psychotherapeutic software lies in the research benefits of fully specified psychotherapeutic interventions. Among the most striking methodological improvements in psychotherapy research over the past 25 years has been the introduction of treatment manuals to specify interventions under study. These are used both in the training of therapists taking part in research studies and as the basis for rating systems applied to recordings or transcripts of therapy sessions in order to establish the integrity of the treatment as delivered within the study. Thus, manuals serve to reduce unaccountable variations in what the therapist does during treatment sessions, increasing the confidence with which we can claim that the treatment was delivered in accordance with its specification. Accordingly, the use of manuals has become commonplace in studies of the efficacy or effectiveness of psychotherapies.

However, all the actions of therapists, their trainers and the raters tasked with assessing their adherence or competence are human actions vulnerable to variation and error, so that the specification achieved via manualization is at best partial rather than complete. This variation is at best random (increasing 'noise' and thus reducing the power of a study to detect the differences between treatment groups sought by outcome research) and at worst systematic (introducing bias into the study). For example, the results of the NIMH Treatment of Depression Collaborative Research Program (TDCRP) (Elkin, 1994) were disappointing to proponents of cognitive-behavioural therapy and have been attributed to variations across the participating treatment sites in the quality with which treatments were delivered (Jacobson and Hollon, 1996).

In contrast, consider a computer program. Once written and implemented, provided all bugs have been detected and eliminated, it performs exactly as specified. This enables more powerful studies that are less vulnerable to bias. The computerization of psychotherapies may be the next evolutionary step from manualization. That is not to deny, of course, that human error or variation can influence such factors as the confidence instilled into the client concerning the value of using the program. But the integrity of the program itself is assured.

CONCLUSION

To date the development of computerized psychotherapies has been characterized by attempts to imitate present-day human performance in this domain. Are human and computer-conducted psychotherapeutic dialogue comparable in scope and helpfulness? Does computer-assisted dilemma-solving match that performed by guidance counsellors? Are cognitive and behavioural techniques proven beneficial when delivered by human therapists equally beneficial when offered by a computer?

However, perhaps it is now time to end the competition. We need to establish how empirically supported, computerized psychotherapies can be used stand-alone, adjunctively or sequentially with human experts to offer the best possible care for people who can benefit. We have in our midst available and accessible computerized psychotherapies designed to address unmet population needs. A growing body of evidence supports their clinical benefits and likely cost-effectiveness as a treatment option. Is it unethical to ignore these tools and withhold treatment from those who could benefit?

ACKNOWLEDGEMENT

All images in this chapter have been provided courtesy of Ultrasis plc.

REFERENCES

Australian Institute of Health & Welfare (2002) http://www.aihw.gov.au/media/-2002/mr020531.pdf.

Bebbington, P.E., Brugha, T.S., Meltzer, H., Jenkins, R., Ceresa, C., Farrell, M. and Lewis, G. (2000a) 'Neurotic disorders and the receipt of psychiatric treatment', *Psychological Medicine*, **30**: 1369–76.

Bebbington, P.E., Meltzer, H., Brugha, T.S., Farrell, M., Jenkins, R., Ceresa, C. and Lewis, G. (2000b) 'Unequal access and unmet need: Neurotic disorders and the use of primary care services', *Psychological Medicine*, **30**: 1359–67.

Beck, A.T., Freeman, A., Pretzer, J., Davis, D.D., Fleming, B., Ottaviani, R., Beck, J., Simon, K.M., Padesky, C., Meyer, J. and Trexler, L. (1990) *Cognitive Therapy of Personality Disorders*. Guilford, New York.

Clark, D.C. and Fawcett, J. (1992) 'Review of empirical risk factors for evaluation of the suicidal patient'. In Bongar, B.M. (ed.) *Suicide: Guidelines for Assessment, Management, and Treatment*. Oxford University Press, Oxford, pp. 16–48.

Colby, K.M., Faught, W.S. and Parkison, R.C. (1979) 'Cognitive therapy of paranoid conditions: Heuristic suggestions based on a computer simulation model' *Cognitive Therapy and Research*, **3**: 55–60.

Department of Health (2001) *Treatment Choice in Psychological Therapies and Counselling: Evidence Based Clinical Practice Guidelines*. Department of Health, London.

Elkin, I. (1994) 'The NIMH Treatment of Depression Collaborative Research Program: Where We Began and Where We Are.' In Bergin, A.E. and Garfield, S.L. (eds) *Handbook of Psychotherapy and Behaviour Change*, 4th edn. Wiley, New York, pp. 114–39.

Erdman, H.P., Klein, M.H. and Greist, J.H. (1985) 'Direct patient computer interviewing', *Journal of Consulting and Clinical Psychology*, **53**: 760–73.

Fowler, R.C., Rich, C.L. and Young, D. (1986) 'San Diego suicide study: II. Substance abuse in young cases', *Archives of General Psychiatry*, **43**: 962–5.

Frances, A. & Clarkin, J.F. (1981) 'No treatment as the prescription of choice', *Archives of General Psychiatry*, **38**: 542–5.

Gelso, C. and Hayes, J. (eds) (1998) *The Psychotherapy Relationship: Theory, Research and Practice*. John Wiley, Chichester.

Ghosh, A. and Marks, I.M. (1987) 'Self-treatment of agoraphobia by exposure', *Behaviour Therapy*, **18**: 3–18.

Goldberg, D. and Huxley, P.J. (1997) *Common Mental Disorders: A Bio-social Model*, Routledge, London.

Graham, C., Franses, A., Kenwright, M. and Marks, I.M. (2001) 'Problem severity in people using alternative therapies for anxiety difficulties', *Psychiatric Bulletin*, **25**: 12–14.

Greist, J.J., Gustafson, D.H., Strauss, F.F. et al. (1973). 'A Computer Interview for Suicide-Risk Prediction.' *American Journal of Psychiatry*, **130**: 1327–32.

Haaga, D.A.F. (2000) 'Introduction to the special section on stepped care models in psychotherapy', *Journal of Consulting and Clinical Psychology*, **68**: 547–8.

Harris, E.C. and Barraclough, B. (1997) 'Suicide as an outcome for mental disorders: A meta-analysis', *British Journal of Psychiatry*, **170**: 205–28.

Hassan, A.A.M. (1992) Comparison of computer-based symbolic modelling and conventional methods in treatment of spider phobia. Doctoral dissertation, University of Leeds.

Henriksson, M.M., Aro, H.M., Marttunen, M.J. and Heikkinen, M.E. (1993) 'Mental disorders and comorbidity in suicide', *American Journal of Psychiatry*, **150**: 935–40.

Horvath, A. and Luborsky, L. (1993) 'The role of the therapeutic alliance in psychotherapy'. *Journal of Consulting and Clinical Psychology*, **61**:561–73.

Jacobson, N.S. and Hollon, S.D. (1996) 'Prospects for future comparisons between drugs and psychotherapy: lessons from the CBT-versus-pharmacotherapy exchange', *Journal of Consulting and Clinical Psychology*, **64**: 104–8.

Joinson, A. (1998) 'Causes and implications of disinhibited behaviour on the Internet', In J. Gackenbach (ed.) *Psychology and the Internet: Intrapersonal, Interpersonal, and Transpersonal Implications*. Academic Press, San Diego, pp. 43–60.

Kenwright, M., Liness, S. and Marks, I.M. (2001) 'Reducing demands on clinicians time by offering computer-aided self help for phobia/panic: feasibility study', *British Journal of Psychiatry*, **179**: 456–9.

Lang, P.J., Melamed, B.G. and Hart, J. (1970) 'A psychophysiological analysis of fear modification using an automated desensitisation procedure', *Journal of Abnormal Psychology*, **76**: 220–34.

Lovell, K. and Richards, D. (2000) 'Multiple access points and levels of entry (MAPLE): ensuring choice, accessibility and equity for CBT services', *Behavioural and Cognitive Psychotherapy*, **28**: 379–92.

Maris, R.W., Berman, A.L., Maltsberger, J.T. and Yufit, R.I. (eds) (1992) *Assessment and Prediction of Suicide*. Guilford Press, New York.

Marks, I., Shaw, S. and Parkin, R. (1998) Computer-aided treatments of mental health problems. *Clinical Psychology: Science and Practice*, **5**: 151–70.

McNeil, B., May, R. and Lee, B. (1987) 'Perceptions of counsellor source characteristics by premature and successful terminators', *Journal of Counselling Psychology*, **34**: 86–9.

Mead, N. and Bower, P. (2000) 'Patient-centredness: A conceptual framework and review of the empirical literature', *Social Science and Medicine*, **51**: 1087–110.

Meltzoff, J. and Kornreich, M. (1970) *Research in Psychotherapy*. Atherton Press, New York.

Melzer, H., Gill. B., Petticrew M. and Hinds K. (1995) *OPCS Surveys of Psychiatric Morbidity in Great Britain. Report 1. The Prevalence of Psychiatric Morbidity among Adults Living in Private Households*. London: HMSO.

National Institute for Mental Health (1998) http://www.nimh.nih.gov/anxiety/adfacts.cfm.

Norcross, J.C., Santrock, J.W., Campbell, L.F., Smith, T.P., Sommer, R. and Zuckerman, E.L. (2000) *Authoritative Guide to Self-help Resources in Mental Health*. Guilford Press, New York.

Ohayon, M.M., Priest, R.G., Guilleminault, C. and Caulet, M. (1999) 'The prevalence of depressive disorders in the United Kingdom', *Biological Psychiatry*, **45**: 300–7.

Orlinsky, D.E., Grawe, K. and Parks, B.K. (1994) 'Process and outcome in psychotherapy – noch einmal'. In Bergin, A.E. and Garfield, S.L. (eds) *Handbook of Psychotherapy and Behaviour Change*, 4th Edn, pp. 270–376, John Wiley, New York.

Peck, D. (2002) 'Computers, clinicians, conundrums and controversial conclusions', *Proceedings of The British Psychological Society*, **10**(2): 92.

Pope, K.S. (1990) 'Therapist-patient sex as sex abuse: Six scientific, professional, and practical dilemmas in addressing victimization and rehabilitation', *Professional Psychology: Research and Practice*, **21**: 227–39.

Proudfoot, J., Goldberg, D., Mann, A., Everitt, B., Marks, I. and Gray, J. (2003a) Computerised, interactive, multimedia cognitive behavioural program for anxiety and depression in general practice. *Psychological Medicine*, **33**(2): 217–27.

Proudfoot, J., Swain, S., Widmer, S., Watkins, E., Goldberg, D., Marks, I., Mann, A & Gray., J.A. (2003b) 'The development and beta-test of a computer-therapy program for anxiety and depression: hurdles and preliminary outcomes'. *Computers in Human Behaviour*, **19**(3): 277–89.

Robins, L.N. and Regier, D.A. (eds) (1990) *Psychiatric Disorders in America, The Epidemiologic Catchment Area Study*. Free Press, New York.

Roth, A. and Fonagy, P. (1996) *What Works for Whom: A Critical Review of Psychotherapy Research*. Guilford Press, London.

Selmi, P.M., Klein, M.H., Greist, J.H., Sorrell, S.P. and Erdman, H.P. (1990) Computer-administered cognitive-behavioral therapy for depression, *American Journal of Psychiatry*, **147**: 51–6.

Shapiro, D.A. and Cavanagh, K. (2002) It's Not Fair: Inequitable Availability of Cognitive Behavioural Therapy in England and Wales. Unpublished manuscript.

Snyder, C.R., Michael, S.T. and Cheavens, J.S. (1999) 'Hope as a psychotherapeutic foundation of common factors, placebos, and expectancies' In Hubble, M.A. and Duncan, B.L. (eds) *The Heart and Soul of Change: What Works in Therapy*, American Psychological Association, Washington, pp. 179–200.

Spero, M.H. (1978) 'Thoughts on computerised psychotherapy', *Psychiatry: Journal for the Study of Interpersonal Processes*, **41**: 279–88.

Stiles, W.B. and Shapiro, D.A. (1989) 'Abuse of the drug metaphor in psychotherapy process-outcome research', *Clinical Psychology Review*, **9**: 521–43.

Waddington, L. (2002) 'The therapy relationship in cognitive therapy: a review', *Behavioural and Cognitive Psychotherapy*, **30**: 179–91.

Wagman, M. (1980) 'PLATO DCS: An interactive computer system for personal counseling', *Journal of Counseling Psychology*, **27**: 16–30.

Wagman, M. and Kerber, K.W. (1984) 'PLATO DCS, an interactive computer system for personal counseling: further development and evaluation'. *Journal of Counseling and Clinical Psychology*, **27**: 31–9.

Wampold, B.E. (2001) *The Great Psychotherapy Debate: Models, Methods, and Findings*. Lawrence Erlbaum, Mahwah, NJ.

Watkins, E. & Williams, R. (1998) 'The efficacy of cognitive behavioural therapy', *Cognitive Behaviour Therapy*, **8**: 165–87.

Wright, J.H., Wright, A.S., Salmon, P., Beck, A.T., Kuykendall, J., Goldsmith, L.J. and Zickel,

M.B. (2002) 'Development and initial testing of a multimedia program for computer-assisted cognitive therapy', *American Journal of Psychotherapy*, **56**: 76–86.

Zack, J.S. and Hill, C. (2001) *The Dream Toolbox: An Interactive Program for Working with Dreams*. Computer Software. Coral Gables: Authors.

Zarr, M.L. (1984) Computer-mediated psychotherapy: Toward patient-selection guidelines', *American Journal of Psychotherapy*, **37**: 47–62.

Conclusion

KATE ANTHONY AND STEPHEN GOSS

During the course of this book, we have seen a number of ways in which technology has started to become part of the overall environment of the mental health profession. It is impossible to predict when or even if some of the applications described here will become such commonplace methods of communication as to no longer be thought of as remarkable 'technological' innovations when used to deliver counselling or psychotherapeutic services. Many would argue that the use of email and attachments have already become as unremarkable for other uses as the telephone. Indeed, all the material in this book was delivered from around the world by electronic attachments, and much of the editing process and development was done via email rather than by telephone or in face-to-face meetings. It is probable that access to broadband Internet will render remote communication the norm rather than the exception for international collaboration between mental health professionals, where the possibility of face-to-face communication is limited. It is predicted that 95 per cent of people in advanced nations will be computer literate by 2010 (BT Exact, 2001) and it is already suggested that the term 'newbie' (coined in 1982 by Martha Ainsworth to describe someone new on the Internet) is becoming redundant (Fenichel, 2002).

We have focused on three main uses of technology for professional interaction and client interaction: Internet text applications (email and IRC), telephone/video technology and stand-alone software. Each has their advantages and disadvantages. To a greater or lesser extent, it is possible to argue that each constitutes a suitable technology for application within mental health. Clearly, there is a need for much further research on understanding the process of the communication and the effects that the technological applications have on the quality of the therapeutic or professional interaction, even in the case of the technology longest established in therapy provision: the telephone.

COMMON ADVANTAGES AND LIMITATIONS

We can distinguish between two uses of technology to conduct a relationship; the first being *practical* in addressing environments where access to face-to-face therapeutic assistance is somehow limited, and the second being *psychological* in that the client may prefer not to sit in the same room as his or her practitioner, finding that distance modalities have advantages over more traditional methods.

The logistical reasons for using distance methods for communication in mental health can be generalized across many of the technological applications discussed. Perhaps one of the most obvious is that access to practitioners and clients is now possible globally, and the need for travel to sessions or meetings is greatly diminished. This is true not just when the entire course of therapy is conducted remotely, but also where the client's absence would mean a gap or break in the process of face-to-face therapy. Indeed, minimizing the impact of enforced breaks in a client's ongoing and otherwise face-to-face therapy has provided the impetus behind several explorations of technology as a means by which distance can become less problematic (for example Lago et al., 1999). For geographically remote areas, the development of technological systems of service delivery will undoubtedly impact on communities previously unable to access therapeutic assistance. The benefits of a cost-effective service delivered and received to and from a home or work office, particularly with regard to training and supervision, will often outweigh the initial outlay for hardware. In addition, losing the need for travel means that the practitioner's and client's time constraints are made less severe. Asynchronous methods of service delivery also mean that global time zones are reduced as a barrier to services.

Perhaps the second most obvious benefit of these newly accessible services is for those with physical disabilities, rendering travel to therapy difficult or impossible. With physical disability, the technology goes much further than considerations of just access – hardware and software can be adapted to assist the use of computers, telephones and videoconference services, as noted in Chapter 1. Treatment of psychological conditions such as agoraphobia or social anxiety can also be addressed whilst allowing clients to receive therapy in the home, again giving reason to expect that access to much needed services could be improved.

Cultural differences across the world are also a reason why clients and practitioners may choose remote ways of working. People living in a culture other than their own may feel more comfortable in having access to someone who is based in their country of origin with more understanding and experience of their issues. In addition, the globalization of mental health may reduce the stigma of seeking therapy in itself, as more information is made more freely available. Information about what therapy is and what it can achieve is easily posted on websites, with many Internet and software facilities available to translate, often automatically, into many languages.

Another practical consideration for all these services (and also face-to-face methods) is the improved ability to communicate previously paper-based items, such as a client's homework journal or an information leaflet about a condition, via electronic services such as email or fax. This can be almost instantaneous and greatly reduce the need to spend session time in conveying information, particularly technical specifications and instructions. In the USA, Childress and Williams (2002) use these devices for patients with schizophrenia to self-monitor their symptoms, allowing for early intervention in the event of deterioration or breakdown. Others have used computers to assist in appropriate intervention-planning (Oher et al., 1998; Kok and Jongsma, 1998; Mayers and Rabatin, 2000; Rich, 2001).

There are also the psychological reasons for a client preferring to use distance methods of therapy. It has been stated, and is worth reiterating, that sometimes, whether we as professionals like it or not, clients prefer not to sit with us. Professor Dave Mearns (2002, p. 15) recounts his own experience of this:

> My view was that Internet counselling, in missing the face-to-face contact (except in videoconferencing), would lose the powerful relational quality of therapy. Windy [Dryden], as is his wont, scorned my naivety and observed: 'Do you...have the temerity to imagine that the Internet nerd would come within 10 miles of your fancy office. No, if you are going to relate with him, you have to go to meet him in his territory – the Internet!'

The term 'nerd' is often used in a derogatory way, and is of course a far from accurate description for many who might feel the same way about a therapist's office, but the point is well made. Many people of all kinds are simply not so keen on spending time in a practitioner's consulting room than the practitioner might wish. Shame, embarrassment, anxiety, self-consciousness or simply being nervous are all reasons why a client may not feel able to face another human being while recounting personal and distressing material and why clients will choose a less physical method of interacting with a therapist. As was stressed in Chapter 1, there are varying degrees of physical presence that can be achieved with technological interfaces. They range from the absence of the practitioner altogether as in the case of some stand-alone software, through text-based media such as email and Internet Relay Chat, to video links offering face-to-face interaction but from a distance, facilitating a sense of less space being invaded, more privacy and the client feeling less under scrutiny and freer to behave as they wish (Ross, 2000). Clients are able to access help from a familiar and safe environment, often with their home comforts around them and, in the case of asynchronous communication, at their own time and pace. We have seen that a feeling of being more in control of the therapeutic situation is often reported, with the possibility of logging off, putting the telephone down, moving out of the range of the camera or exiting the software – all, perhaps, easier to do than leaving a consulting room to terminate a session prematurely.

However, there are also limitations in using technology to facilitate therapy. Perhaps the most obvious is that we are reliant on the stability of the hardware and software itself. This is particularly applicable in small operations where minimizing the likelihood of technical breakdown is a cost consideration in what can be afforded by the practitioner in terms of system performance. Technical breakdown is an irritation at best and may cause psychological damage at worst, as well as opening up problems in ensuring adequate client safety during a session when the communication link is unexpectedly severed. In addition, the inadvertent spreading of malicious viruses (for example a disguised programming code that causes some, usually undesirable, event such as start-up problems) and similar pieces of destructive software may all have a detrimental psychological effect on the vulnerable recipient, possibly creating feelings of paranoia or victimization.

Furthermore, it must be acknowledged that, for a proportion of the population, none of the methods described in the preceding pages will ever be appropriate. Use of even unsophisticated technological applications and devices produces anxiety in many people, and while more and more of us are turning to it to aid and facilitate our work, consumer and leisure pursuits, it is worth noting that technostress and technophobia are well-documented conditions (for example Weil and Rosen, 1998; Brosnan, 1998). In addition to negative reactions to the technology itself, we must consider the effects of what is known as 'information overload', now that the act of searching the vast areas of the Internet for information is so easy and the sending of email (and the ability to copy a message to many people) has proliferated to such an extent (Heylighen, 1999). Being available to our clients and each other 24 hours a day via an Internet connection produces boundary issues, such as whether we can ever really switch off from wondering whether someone has tried to email us and how important that missive could be, particularly when anticipating sensitive material from a client in distress. Mobile telephone technology also underlines this sense of feeling open to receiving communications.

An appropriate point for consideration here is also what effect the developing theories of distance therapy and communication have on established modes of working. As it becomes clear, however anecdotally, that it is possible to develop and maintain a therapeutic relationship via technology, despite the lack of various degrees of verbal and non-verbal cues, the impact of this on our original training in face-to-face work and theoretical orientation may cause us to question some of our basic assumptions about how a relationship between therapist and client exists. In addition, we must consider what would have been possible with a client who is unable to communicate effectively while sitting with us in a consulting room had an alternative method of communication been offered as standard. However, in all the methods of distance therapy considered here, it is obvious that compensatory strategies are usually appropriate to ensure miscommunications and misunderstandings are kept to a minimum.

Technology development moves quickly. As with any innovative area of practice, the pioneers will find that training programmes are unavailable in the early stages of exploration. This has classically been the case with email and IRC, where therapy was offered at least five years before any dedicated training programmes were available. Speculative supportive evidence was in abundance, although more rigorous research was, and still is, largely absent. The ethical provision of training in these methods must rely on the research evidence available to date. At the time of writing, a small number of courses are now available, developed largely by those who had immersed themselves in technique and research and were able to develop courses that they felt appropriate from a theoretical, ethical and practical stance. Issues of regulation have had to be addressed by the online mental health community themselves, creating verification sites in an attempt to offer consumers some sort of indication of the quality of service they could expect.

Remote services, whether synchronous or asynchronous, text-based or tele/video-based, are not for the novice practitioner (Goss et al.,

2001). Experience in practising face-to-face therapy before attempting the use of technology to convey or conduct a therapeutic experience is a necessary prerequisite. Furthermore, it is accepted here that distance training is not suitable for becoming a therapist in itself. The authors do not believe that it is possible to develop adequate counselling or psychotherapeutic skills through distance methods for them to be regarded as adequate to the task. However, it is arguable nonetheless that in order to train for distance therapy, the training itself should be conducted, at least in part, via the technology for which the practitioner is being prepared. This sets up the practitioner with experience not only in the logistics of working with the technological application being used but, more importantly, for the differences and adjustments that need to be made as a result of using it to conduct a therapeutic relationship.

Demonstrating actual client sessions for training purposes in most technologically mediated therapies is often relatively easy, whether through anonymized or fictional emails (both client mail and practitioner response), through programming that enables the scrolling of real-time IRC sessions or recordings of sessions provided for the purpose. In the case of email and IRC, questions and comments can appear as pop-up screens for consideration, with interactive coursework being submitted and saved on the server for access by both trainee and trainer. In addition, role-play triads (client/practitioner/observer) can be conducted live in chat rooms. The use of discussion boards, direct links to research and information websites, and email listservs all mean that training can be enriched by online communication strategies while learning about their use in therapy, without any verbal or face-to-face input, precisely as could be expected in the trainees' future relationships with clients.

THE NEED FOR, AND BARRIERS TO, INTERNATIONAL REGULATION

It is perhaps ironic that one of the greatest incentives for developing technological means of delivering psychological services should lead to one of its most potentially problematic aspects. Services that rely on technology are often developed to address problems of geographical distance (see Chapters 5 and 6). Many, most obviously those that make use of the Internet such as email and IRC services, then have the potential to create a global reach. In creating the opportunity for clients and practitioners to contact each other from anywhere in the world, questions of quality control, professional regulation and client safety become acute. In different parts of the world, the way in which the psychological therapies are overseen, if they are at all, varies enormously.

Even the meaning of the common terms used can be understood very differently in different cultures. Consequently, cultural issues are among those that must be tackled by practitioners offering distance therapy beyond their usual context and range of experience. Some services (such as those relating to marriage guidance, sex, sexuality or abortion) are entirely taboo or otherwise unacceptable in some cultures. It has never been the case that practitioners are required to avoid crossing cultural boundaries. Neither is it the case that being controversial, unpopular or simply out of step with the norms and expectations of any given social context is an indication of poor practice. However, whenever practitioners provide services to clients from a cultural background different from their own, a sufficient degree of cultural sensitivity is requisite. It would be unusual in the extreme, perhaps impossible, for any one practitioner to be able to have a thorough understanding of the cultural context for clients from absolutely anywhere, even just within the English-speaking world. It is therefore advisable to restrict one's practice to those areas in which one is confident that such challenges will not be beyond what might be expected in a normal consulting room. In practice, this consideration alone may restrict the otherwise unfettered internationalism of distance therapy services to those areas to which the service in question is truly transferable.

Cultural issues aside, adequate quality control requires that clients who receive poor service should have a right of redress through an established complaints procedure. The variation in regulatory procedures around the world renders this a potentially complex task unless the practitioner's own professional body is willing to act on complaints from a member regardless of the geographical location of the client. The British Association for Counselling and Psychotherapy (BACP), for example, recommends that 'regardless of the location of their client practitioners should always consider themselves bound to maintain *at least* the standards of practice required by their own professional organisation' (Goss et al., 2001, p. 5, emphasis in the original). This does not, in itself, address the issue of what clients might expect of their practitioner. The BACP guidance goes on to state that where differences in standards exist, such as the level of qualification expected or licensing arrangements, practitioners should assume that *both* sets of requirements apply. Thus, practitioners should consider themselves bound to measure up to the standards of their client's country, as well as their own. However, numerous professional bodies around the world are yet to adopt a code of practice that would cover such issues.

A converse concern for practitioners is that it remains entirely possible that clients could attempt to complain, or even take legal action from within their own country. In some circumstances, the practitioner or service provider may find themselves held to account in a system with which they are unfamiliar. For example, the issues relating to who is considered to be under the age of consent, and the implications of working with them, are particularly sensitive. However, reciprocal legal arrangements between countries vary enormously. For example, the USA and UK, politically and economically allied as they are, do not have a mutual extradition treaty. Where they cannot be compelled to participate in proceedings against them, a practitioner might decide not to cooperate. The least result might be that travel to the client's country could become difficult but even this is not certain and, in any case, it might be argued that ethical obligations override legal ones in such circumstances.

The implications of these complexities, in the absence of any generally agreed, internationally applicable ethical standards, is that practitioners and clients who operate on the global scale that the Internet appears to promise are exposed to a degree of risk that would not be present if they remained within their own legislative area.

THE NEED FOR RESEARCH

At the time of writing, technologically mediated therapy of all types remains a seriously underresearched area. Counselling and psychotherapy are evidence-based professions and it is vital that all our activities, most especially in innovative areas, progress only *with* the evidence, not ahead of it. At present it is not possible to determine the comparative effectiveness of traditional face-to-face sessions with most of the methods described above, with the possible partial exception of some of the examples of the stand-alone software products described in Chapters 8 and 9. In the majority of instances, the research is either not yet done or is still at the exploratory stage.

As mentioned above, it is noticeable that this is perhaps most true of the longest established technologies such as the telephone. In contrast, the research base alluded to in Chapter 6 demonstrates the much greater level of research activity that has gone into exploring the telephone's more modern counterpart: videoconferencing. Email and IRC-based services are readily available in many parts of the world and yet they are still to be subjected to rigorous research as generally understood by the profession at large (Rowland et al., 2000; Goss and Rose, 2002).

The most important area, and the one that perhaps requires the most urgent attention, is the need to establish the effectiveness, efficacy (Roth and Fonagy, 1996) and safety of online provision. It could be argued that those who are prepared to offer services to the public ahead of definitive research on these matters are acting prematurely, despite the positive accounts given in the preceding chapters. Thus far, again with the exception of therapeutic software products, most research has been limited to relatively small-scale enquiries rarely even attempting to address such issues. This research need not necessarily be expensive or large-scale, although there is also a clear need for randomized controlled trials that generally require significant levels of funding in themselves. Single case studies can, if sufficiently rigorous and thorough, convincingly demonstrate cause and effect between intervention and outcome, while at the same time providing detailed information on the experience of therapy from the perspectives of both client and therapist (for example Elliott, 2002).

However, alongside research relating to outcomes, it is also imperative that much further work, under carefully controlled, safe conditions, is carried out into the processes involved in therapy when provided via a technological medium. The subject of the human–computer interface is itself a rich area for investigation, already creating a large literature, albeit mostly restricted to more everyday applications. The very fact that providing mental health services via technology is controversial among practitioners is itself an argument for using this emerging field as a basis for testing the extent to which new technologies can be used to facilitate therapeutic communication and the degree to which they can be substituted for meeting one another face to face. The psychological therapies depend, for the most part, on a delicate, intricate and often highly emotionally charged interaction between client and practitioner. If a given kind of technologically mediated communication can be a means by which counselling and psychotherapy can be carried out effectively, then it must be fairly certain that it is possible to use it for most other purposes. That is, if the subtleties of communication and nuances of meaning and significance on which so much of the psychological therapies depend can be effectively maintained or reproduced, there is little more that could be asked in any other application.

Among other topics that are ripe for research attention, we must identify which client groups are most, and least, likely to benefit from these kinds of therapeutic provision. Without this basic information, it is difficult for a practitioner to make a sufficiently thorough assessment of their potential client and decide whether this or that individual should be accepted for the intervention on offer, or whether their interests are better served by referring them on to different kinds of intervention.

Many groups have been proposed as particularly unsuitable for distance therapy provision. Examples include those who are suicidal or under the age of consent. The suicidal, it is suggested, need a greater level of assistance than is typically possible at a distance and may need to be protected from self-harming by direct intervention. Working with those who are under the age of consent may be problematic for practitioners, especially if consent from a parent or guardian is not forthcoming or verifiable. Conversely, however, for most such groups, it is also arguable that online services may be especially appropriate. For example, although not counselling or therapy provision per se, in the UK the Samaritans have offered an email service explicitly targeted at the suicidal since 1994 and report favourably both on its success and rapidly increased use (Lago et al., 1999). It is often thought to be difficult to encourage young people to use psychological services and yet many are likely to be more at home with online communication than their elders. The difficulty of defining who is suitable, or otherwise, for online therapy is exemplified by the conflict between the sensible caution regarding international service provision noted above, and the immense opportunity offered by the Internet to render mental health services far more equitably accessible by cultural and national groups who are otherwise routinely underrepresented among our clients.

It is also necessary that we investigate the processes of online therapy in order to identify them, develop their strengths and prevent, or limit, the risks. Furthermore, we must explore the boundaries and limitations of what can be done as well as how we can best work within them. It is possible that we may be able to achieve at a distance things that cannot be achieved in other ways – or that the different technologies each have their own limits that are quite different from other methods. It may be, as suggested by many of the contributions to this book, that similar, or equivalent, phenomena can be created in technologically mediated therapeutic relationships as when working face to face. Such lines of investigation will be especially crucial in facilitating the conversion of practitioner skills from face-to-face settings to cyberspace. They will also be essential for online services to accurately represent themselves to potential clients.

Despite the increasing body of theoretical and anecdotal literature on the subject, many aspects of technology in mental health provision are yet to have any definitive empirical basis and, as a result, all the optimism shown by its proponents must be subject to caveat. The vast amount of work that remains to be done should be a strong stimulus to research activity of all kinds and, at least in some areas of importance, studies are already underway. We will have to wait for convergent findings on each new technology, as it appears, to create an empirically supported evidence base before we attempt less speculative and more definitive comment.

A LOOK TO THE FUTURE

The concept of using even more innovative technology for mental health provision is already being realized. Virtual reality environment applications are becoming recognized as a valuable way of studying cognitive and behavioural processes, and can be designed to record data and measure effectiveness automatically via the apparatus used, such as a head-mounted device worn to experience visual three-dimensional and audio input. Examples of this are the creation of a three-dimensional classroom with different distraction stimuli to study the concentration of children with ADHD and assess their needs (Rizzo et al., 2001), and the positive use of a therapeutic virtual environment for adolescents facing haemodialysis as a way for patients to distance themselves from the harshness of their medical condition (Bers et al., 2001). Further examples include the combination of a virtual environment and video footage to recreate and address anxiety-provoking situations such as public speaking (see www.virtuallybetter.com), where the video audience can be controlled from a keyboard to react appropriately to the speaker, including clapping, laughing, walking out of the auditorium or throwing balls of paper. Social phobias can be examined and treated by exposure to virtual parties, and acrophobia dealt with similarly by use of a virtual lift (Hodges et al., 1995) to desensitize clients to their fear (see also Chapters 8 and 9). NASA is able to examine the effects of an isolated work environment such as a space station (see http://spaceflight.nasa.gov/) and is experimenting with recreating a human presence by virtual reality to examine and avoid psychological breakdown due to isolation in those environments. Haptic bodysuits that allow people to occupy the same virtual space while many miles apart will offer total immersion in a virtual environment with tactile feedback, meaning that therapy can take place as an actual face-to-face session remotely. One of the advantages of these sorts of study is the ability to expose several subjects to the same stimuli, allowing for comparative data in a controlled environment.

In addition to devices such as head-mounted apparatus for recording data, the use of clothing containing sensors to monitor physiological changes can assess what other bodily reactions exist in times of trauma. For example, when used in conjunction with virtual reality systems to recreate scenes from the Vietnam War to treat soldiers with post-traumatic stress disorder (see Chapter 8 and www.vivometrics.com for examples such as the *LifeShirt* system), such systems allow the study of psychological processes in direct comparison to physiological ones, giving a constant record or 'movie' of both, rather than a 'snapshot' received from a single test of, for instance, heart rate or blood pressure. This can be used to measure therapeutic progress in reactions to anxiety-provoking stimuli as the client becomes desensitized, witnessing (and assessing) his or her own physiological reactions and controlling breathing, for example, in the event of panic.

The development of sophisticated computer-generated representations of the self (avatars) for use in mental health environments is already underway at the time of writing (Anthony and Lawson, 2002). Avatars come in many shapes and forms and in the past their appearance has been limited by the graphical processing power of computer technology. As technology improves, however, the concept of virtual 'avatar therapy' becomes more viable (Goss and Anthony, 2002). Development of these sorts of method is based on the premise that some clients find the idea of anonymous, safe, comfortable and readily available service provision empowering and preferable to traditional (usually face-to-face) therapy. Once created, a remote therapist can control the avatar in a real-time, online therapy session, speaking through the avatar to the client. An intriguing research idea might even be to discover whether clients would find benefit in being able to choose the appearance of their therapist rather than the reality of how they actually look, perhaps choosing, say, a black male therapist instead of a white female one. Testing whether a therapist's physical appearance can become redundant is a luxury technology will give us. Full-body avatars, whether actual representations of the person's physical appearance or not, can mean that groups can meet from anywhere in the world and hold a session in complete anonymity and genuine safety. Put these avatars into specific relaxing virtual environments, and we can lose the impersonal waiting room and consulting room altogether.

Avatars can also be configured to act as a virtual host on a website or CD-ROM, providing 'human-style' interfaces to an automated therapy information system. In this case, the avatar acts independently and uses a natural language engine and knowledge database so that it provides virtual intelligence and knowledge to the client. The use of natural language scripts provides a mental health information service that creates a realistic question and answer session that appears to be a dialogue between two people, when in fact the client is interacting with a piece of software.

The acceleration of technological development is constantly bringing new ideas to the profession of counselling and psychotherapy in how it can be utilized to provide a more efficient and accessible way of gaining mental health assistance and professional collaboration. It is an exciting area but we must not lose sight of the fact that it is improved communication that is the key to achieving this, not the method by which it is done. Technology facilitates communication but does not create it. We may not be able to predict the ways in which the talking therapies will change as a result of using technology, but we can state that it is the *talking* aspect of them that remains, however that communication is passed between client and practitioner. It is our responsibility to provide technological services, where appropriate, in a professional, responsible and ethical manner. By ensuring that we have the research base that affords us the chance to explore how technology can be used in a positive and beneficial way, we can become increasingly sure of it as a way forward for the profession.

REFERENCES

Anthony, K. and Lawson, M. (2002) The Use of Innovative Avatar and Virtual Environment Technology for Counselling and Psychotherapy. Available at: http://www.kateanthony.co.uk/research.html.

Bers, M., Gonzalez-Heydrich, G. and DeMaso, D. (2001) 'Identity Construction Environments: Supporting a Virtual Therapeutic Community of Pediatric Patients Undergoing Dialysis' In Proceedings of Computer-Human Interaction ACM.

Brosnan, M. (1998) *Technophobia: The Psychological Impact of Information Technology*. Routledge, London.

BT Exact (2001) Technology Timeline. Available at http://www.btexact.com/ideas/whitepapers.

Childress, C. and Williams, J. (2002) 'Incorporating Email Into Patient Care at Community Mental Health Clinics'. Paper given at the 110th American Psychological Association Convention, Chicago.

Elliott, R. (2002) Hermeneutic Single-Case Efficacy Design. *Psychotherapy Research*, **12**(1): 1–21.

Fenichel, M. (2002) '21st Century Psychology: Innovations and Challenges of Internet Facilitated Communication'. Paper given at the 110th American Psychological Association Convention, Chicago.

Goss, S.P. and Anthony, K. (2002) Virtual counsellors – whatever next? *Counselling and Psychotherapy Journal*, **13**(2): 14–15.

Goss, S.P. and Rose, S. (2002) Evidence based practice: a guide for counsellors and psychotherapists. *Counselling and Psychotherapy Research*, **2**(2): 147–51.

Goss, S., Anthony, K., Jamieson, A. and Palmer, S. (2001) *Guidelines for Online Counselling and Psychotherapy*. BACP, Rugby.

Heylighen, F. (1999) 'Change and Information Overload: Negative Effects'. Available at http://pespmc1.vub.ac.be/CHINNEG.html [accessed 14 July 2002].

Hodges, L.F., Rothbaum, B.O., Kooper, R., Opdyke, D., Meyer, T., North, M., de Graff, J.J. and Williford, J. (1995) 'Virtual environments for exposure therapy'. *IEEE Computer Journal*, **7**: 27–34.

Kok, J.R. and Jongsma, A.E. (1998) *The Pastoral Counseling Treatment Planner*, John Wiley, New York.

Lago, C., Baughan, R., Copinger-Binns, P., Brice, A., Caleb, R., Goss, S. and Lindeman, P. (1999) *Counselling Online Opportunities and Risks in Counselling Clients via the Internet*. British Association for Counselling and Psychotherapy, Rugby.

Mayers, L.B. and Rabatin, D.L. (2000) *Brief Employee Assistance Homework Planner*, John Wiley, New York.

Mearns, D. (2002) 'New Blood', *Counselling and Psychotherapy Journal*, **13**(5): 14–15.

Oher, J.M., Conti, D.J. and Jongsma, A.E. (1998) *The Employee Assistance Treatment Planner*, Wiley, New York.

Rich, P. (2001) *Grief Counseling Homework Planner*, Wiley, New York.

Rizzo, A.A., Buckwalter, J.G, Humphrey, L, van der Zaag, A., Bowerly, T., Chua, C., Neumann, U. and Kyriakakis, C. (2001) 'A Virtual Reality Classroom Scenario for ADHD Assessment'. The 28th Annual Conference of the International Neuropsychological Society, Denver, CO.

Ross, C. (2000) 'Counseling by Videoconference'. Paper presented at the 5th International Conference on Client-Centred and Experiential Psychotherapy, Chicago, June.

Roth, A. and Fonagy, P. (1996) *What Works for Whom?: A Critical Review of Psychotherapy Research*, Guilford Press, New York.

Rowland, N. and Goss, S. (2000) *Evidence-based Counselling and Psychological Therapies: Research and Applications*, Routledge, London.

Weil, M.M. and Rosen, L.D (1998) *Technostress: Coping with Technology @ Home, @ Work, @ Play*. John Wiley, New York.

Index

I

J

K

L

M

P

Q

R

S